Universe of Particles

A theory of physics

by Fredrik Nygaard (2018)

Introduction

A year has passed since I published two short books on physics in which I used a strict particle model to explain the physical world we live in. In this model, nothing happens without direct physical interaction. There is no action at a distance. Everything that exists is an assembly of particles. Every force is due to particles colliding and interacting.

The books, titled The Velcro Universe and The Velcro Cosmos, outline a physical model that can be used to explain a whole range of physical phenomena. The Velcro Universe covers, electricity, magnetism, gravity, optics, and the atom. The Velcro Cosmos covers space, time, inertia and energy.

Both books were hastily written. They contain no elaborations and no excuses. The physics is laid out without explaining the wider context. My aim was solely to show that it would work, and as such the two books are still relevant. However, there are some errors and misrepresentations. There are also quite a few things that could do with some elaborations.

With the dust now settled from my initial excitement, I feel that time has come to write a revised version of my two books. I will elaborate on my thinking and explain more fully why the model is the way it is, and how it can be used to explain the observed universe.

My aim is not to make the reader a true believer in my theory. I do not believe in "settled science". That goes as much for this theory as any other. Rather than settling on a single theory as being the one and only true representation of reality, I want to see people open up to the idea that the world can be described in many ways, and that all the different ways deserve equal attention and scrutiny.

Settling on a single idea only serves to restrict the mind. As Aristotle once said, it is the mark of an educated mind to entertain a thought without accepting it. The thought I want the reader to entertain throughout this book is the idea that everything we know can be derived from three basic particle quanta, empty space, and a handful of rules to fuse it all together.

Morton Spears' Particle Quanta

In my theory, I use the simple model of the subatomic proposed by Morton Spears in his second book on gravity, published in 1993.

The reason for this is that it is the simplest model I could find that suits the purpose of my physics. Nothing more elaborate is required.

Morton Spears' model is much simpler than standard physicists. Morton Spears does not invoke a large array of Quarks, Leptons and Bosons. Instead he makes the proposition that the subatomic is made up of three indivisible particle quanta. One is negatively charged, the other is positively charged, and the third is neutral.

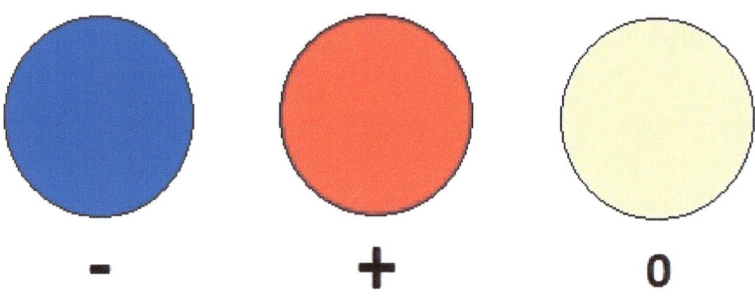

Morton Spears' 3 particle quanta

An objection to this simple alternative to standard theory is that we have evidence to suggest that Quarks, Lepton and Bosons are real. However, the evidence can just as well be used to defend Morton Spears. All that is observed is that when atoms are smashed together with great force, they break apart. A large number of short lived particles appear before promptly recombining into something more stable. Nothing is proven beyond the fact that atoms are made up of a large number of smaller parts.

Choosing conventional particle physics over Morton Spears' model would merely complicate things as far as my particular theory is concerned. I chose therefore to go with Morton Spears, and since nothing turned up in which I would require Quarks, Leptons or Bosons, I stuck with his model.

Morton Spears' Proton, Neutron, Electron and Neutrino

Morton Spears arrived at his particle quanta by comparing the mass of a proton to that of a neutron. The mass of these two particles differ by a ratio of 2177 to 2180. From this, he concluded that a proton is made up of 1089 positive quanta and 1088 negative quanta, and that the neutron is made up of 1090 positive and 1090 negative quanta.

This gives the proton a net charge of 1 and the neutron a net charge of 0.

The total number of charged particles gives us the mass of the proton and the neutron. The net sum of charged particles give us their overall charge.

Since the neutrino has no mass, we do not know how many neutrinos may take part in the construction of a proton or neutron. However, we do know that a free neutron will decay into a proton, an electron and a neutrino in about fifteen minutes when taken out of an atomic nucleus.

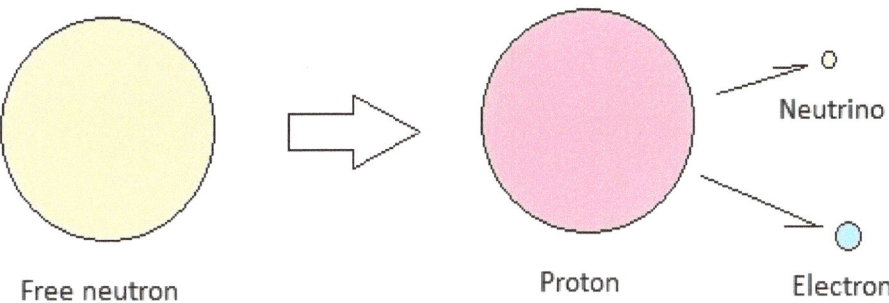

Free neutron decay

This tells us that the difference of 3 quanta between the neutron and the proton must be the charged quanta making up the electron.

The electron is therefore made up of 2 negative quanta and 1 positive quantum. We do not know the role of the neutrino in this. But as far as mass is concerned we know the number of charged quanta making up the neutron, the proton and the electron.

The neutrino is very small. It is therefore reasonable to assume that it is made up of a single neutral quantum.

To summarize, we have the following:

- Proton = 1089 positive quanta + 1088 negative quanta (a total of 2177)
- Neutron = 1090 positive quanta + 1090 negative quanta (a total of 2180)
- Electron = 1 positive quantum + 2 negative quanta (a total of 3)
- Neutrino = 1 neutral quantum

Keeping Things Together

With the exception of the neutrino, all of Morton Spears' particles are composed of three or more particle quanta, and an idea struck me immediately regarding this.

Morton Spears' quanta must have some sort of texture to them so that they can stick together.

Allowing for this, the strong force that holds atomic nuclei together can be explained entirely in terms of texture. The short reach of the strong force corresponds to the short reach of the textures covering each quantum.

Furthermore, if these textures are such that positive quanta are slightly more reactive than negative ones, then the puzzling difference in size between the electron and the proton can be explained. Protons are larger than electrons because positive quanta are a tiny bit more reactive than negative ones.

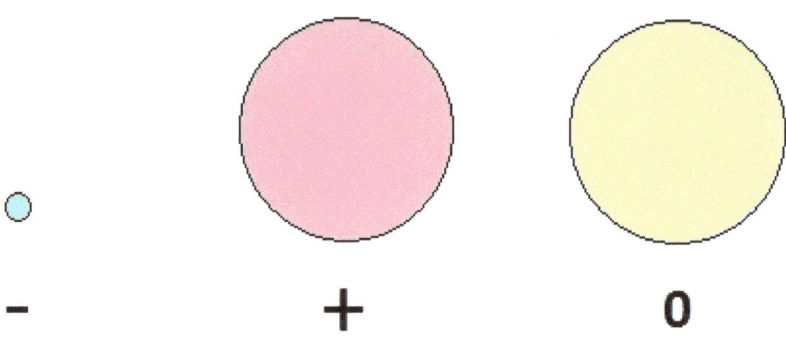

Electron, proton and neutron

For the purpose of illustration, we can use Velcro as an analogy for the two types of textures involved. We can assign hooks to positive quanta and a hoops to negative quanta.

Anyone that has played with Velcro knows that while hoops do not react with other hoops, hooks do react ever so slightly with other hooks. There is a tiny imbalance in reactivity between hooks and hoops.

As for neutral quanta, I came to the conclusion that they would have to have a surface covered in hooks and hoops in equal measures. This would allow them to interact weakly with both electrons and protons.

The point here is not that these quanta are covered in Velcro. The precise nature of the quanta is irrelevant. The point to note is that they have surface features, and that they stick together in ways that are reminiscent of Velcro.

The Dielectric Photon

The electromagnetic phenomena of visible light, radio waves, gamma-rays and the like, are generally ascribed to the photon, a tiny massless particle capable of carrying energy.

Since Morton Spears' neutral quantum is assigned to the neutrino, the photon has to be made of an equal number of positive and negative charged quanta.

The photon has to be dielectric.

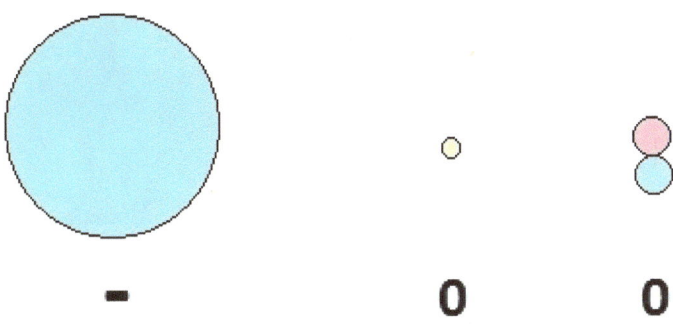

Electron, neutrino and dielectric photon

The fact that photons are much smaller than electrons, and without inertial mass, implies that the quanta involved are in a compact massless state.

For photons to be made of the exact same type of quanta as electrons and protons, Morton Spears' quanta will have to come in two states, one small and compact, and the other large and bloated.

The quanta making up electrons and protons must be in an open state while the quanta making up the photon are in a closed state.

Quanta in the open state, are large, with inertia. Quanta in the closed state are small, with no inertia.

Photons are thus made up of an equal number of positively and negatively charge quanta in the closed state.

As for the exact composition of the photon, it can be derived from the well known phenomenon of electron-positron pair production.

Electron-Positron Pair Production

Gamma-ray photons are known to spontaneously produce electron-positron pairs when in close proximity of massive atomic nuclei. At the exact moment that a gamma-ray disappears, an electron-positron pair appears.

The standard explanation for this is that virtual electron-positron pairs get transformed into real electron-positron pairs by gamma-rays when inside the strong electric fields of massive atomic nuclei.

However, the spontaneous appearance of an electron-positron pair can just as well be explained as a transformation of the photon itself. If the photon is a compact dielectric configuration consisting of a positive orb and a negative orb, we got all the components required to explain the spontaneous appearance of electron-positron pairs.

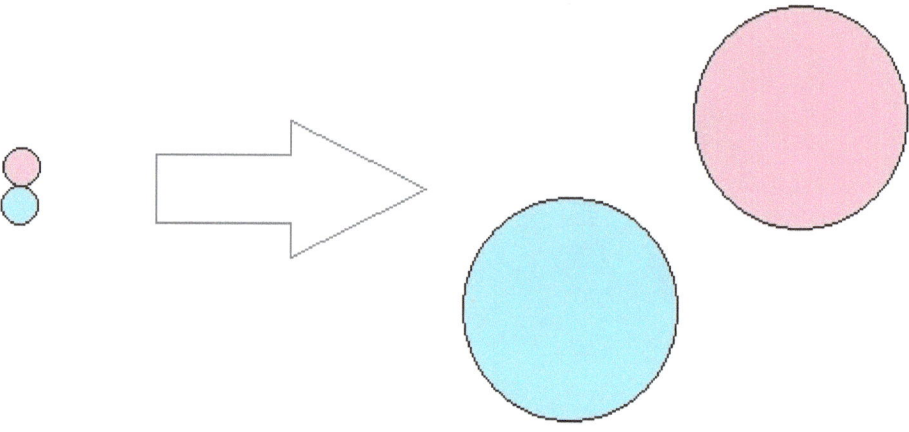

Gamma-ray photon producing an electron-positron pair

We already know that the electron is a configuration of 1 positive quantum and 2 negative quanta.

A positron has the exact same mass as an electron, but with a positive charge. It must therefore consist of 1 negative quantum and 2 positive quanta.

All together we get 3 positive quanta and 3 negative quanta spontaneously appearing from the gamma-ray photon. A photon is therefore constituted of 3 positive and 3 negative quanta.

We can now add the positron and the photon to our list of particles explained entirely in terms of Morton Spears' particle quanta:

- Proton = 2177 charged quanta in the open state (1089 positive and 1088 negative)
- Neutron = 2180 charged quanta in the open state (1090 positive and 1090 negative)
- Electron = 3 charged quanta in the open state (1 positive and 2 negative)
- Positron = 3 charged quanta in the open state (2 positive and 1 negative)
- Neutrino = 1 neutral quantum in the closed state
- Photon = 6 charged quanta in the closed state (3 positive and 3 negative)

Note that the two particles made up of quanta in the closed state both move at the speed of light.

An intriguing conclusion that can be drawn from the above line of reasoning is that photons can be transformed into massive matter through physical manipulations.

When sufficiently stressed, gamma-rays pop like popcorn in a microwave oven. They undergo a phase transition from photon to an electron and a positron.

Conversely, we get that an electron that encounters a positron will spontaneously "annihilate" into a gamma-ray photon. This too is well documented in laboratory experiments. However, with our alternative perspective, nothing disappears. The electron-positron pair is not turned into "pure energy". It is merely popped back into a photon.

Sticky Light

A recent discovery at MIT further supports the idea that photons may be dielectric, and capable of a phase shift into massive matter.

Researchers found that photons can stick together into pairs and triplets when passed through a cloud of ultracold rubidium atoms.

The structures formed had mass, and were therefore slowed down a lot. They moved at a speed 100,000 times slower than ordinary light.

The light appears to have gone through a phase shift, similar to that observed for electron-positron pair production. However, in this case, we got the additional effect of photons latching onto each other to create structure.

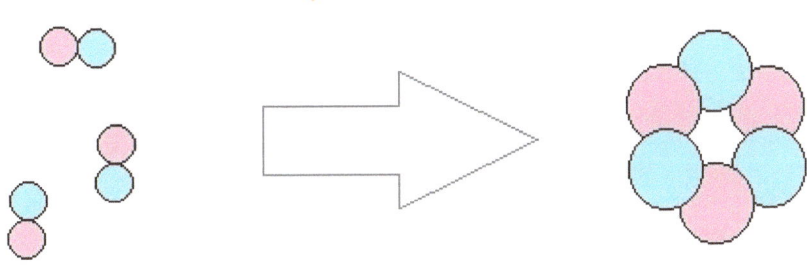

Quasi-stable matter produced from light

Light was, as it were, condensed into matter.

This discovery, made known to me several months after I wrote The Velcro Universe, fits my hypothesis remarkably well.

The discovery that photons can be made to stick together goes a long way towards proving that photons are dielectric. After all, it is in the nature of dielectric matter to interact and form structures.

Four Stable Particles

Of the six particles so far described in terms of Morton Spears' particle quanta, only four are stable. The neutron cannot exist for long outside atomic nuclei, and the positron will quickly find an electron to combine with to produce a photon.

While positrons are highly reactive, combining readily with electrons, neutrons simply fall apart.

The only subatomic particles that are stable enough to exist freely in nature are:

- Proton = 2177 charged quanta in the open state (1089 positive and 1088 negative)
- Electron = 3 charged quanta in the open state (1 positive and 2 negative)
- Photon = 6 charged quanta in the closed state (3 positive and 3 negative)
- Neutrino = 1 neutral quantum in the closed state

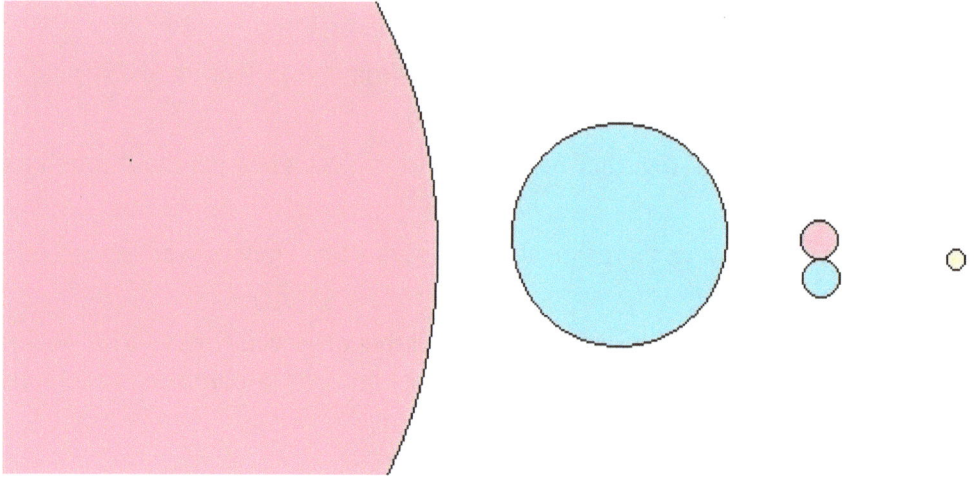

Proton, electron, photon and neutrino

The Velcro Universe does not require a neutron, and does not treat it as a fundamental particle. Instead, it is considered a composite. It is a proton with an electron attached to it.

The fact that the proton is incapable of holding onto the electron is very telling. It indicates that the electric force, supposedly very strong in the close vicinity of a proton, isn't really there. A proton cannot hold onto an electron for much more than fifteen minutes.

A stray electron hitting a lone proton will not produce a neutron. The electron will bounce. If the electron has sufficient energy to escape the pull of the proton, it will disappear into space. If not, the electron will be pulled back down to the proton for another bounce.

The Bouncing Electron

Assuming that things are perfectly elastic at the subatomic level, an electron incapable of escaping the electric field of a proton will not come to rest at its surface. It will bounce. With no energy added to the electron, it will be stuck in limbo. It will neither combine with the proton to form a neutron nor escape into space.

The electron clouds that surround atomic nuclei are manifestations of this. Every cloud corresponds to a bouncing electron.

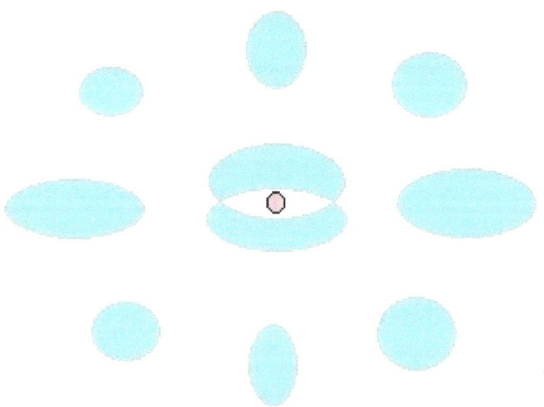

Atomic nucleus with net charge of 10, surrounded by 10 bouncing electrons = Neon

For atoms with more than two protons in their nuclei, there is not enough room for all of the electrons to bounce directly off the nucleus. Only two electrons can do this. Additional electrons bounce off of the electric field of the electrons closer to the nucleus. These electrons are attracted by the nucleus, but repelled by their fellow electrons.

The inner electrons are not free to bounce at any random frequency. They have to bounce at a frequency that resonates with the nucleus. Only specific harmonics are allowed.

The electrons farther out are in turn bound by the frequencies of the electrons closer to the nucleus. All the electrons are therefore bound directly or indirectly to the resonant frequencies allowed by the nucleus.

This is why electron clouds come in a limited number of allowed energy states.

Chemistry

Chemical bonds are fairly easy to understand in terms of electron clouds.

We know that electron clouds around atomic nuclei come in layers, in which the innermost layer can have a maximum of 2, the next one out a maximum of 8, and farther out still another 8, etc.

The way this works when modelled with bouncing electrons is that electrons sometimes find ways to bounce off more than one atomic nucleus at a time.

In the case of hydrogen molecules, we have two protons held together with two electron clouds.

Since every hydrogen atom has 2 available slots in its inner layer, yet only 1 electron cloud, the most efficient configuration of two hydrogen atoms, is to have them share their respective electron clouds so as to make the most of the available space. The electrons bounce alternately off one and the other proton.

The 2 slots available for electrons to bounce are thus filled. Instead of each proton having only one electron bouncing off of them, both of them get two slots filled through mutual sharing of their single electron.

This yields a more efficient configuration, and energy is released in the process.

However, the hydrogen molecule is not particularly efficient. A more efficient configuration can be achieved by combining two hydrogen molecules with a carbon atom. The product of such a configuration yields methane.

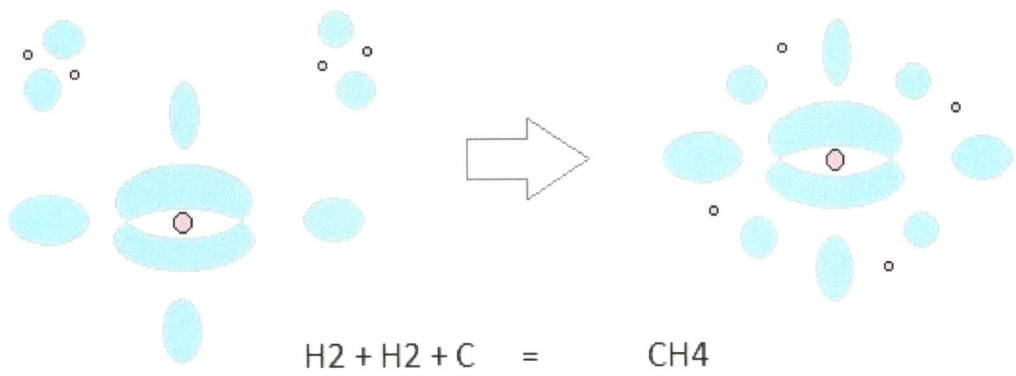

Two hydrogen molecules and a carbon atom combines to form methane

Carbon has 4 empty slots in its second layer. When these slots get filled with electron clouds associated with the hydrogen molecules, an efficient configuration can be made in which all 8 outer slots of the carbon atom are filled in such a way that every hydrogen atom has its 2 slots occupied.

Note that efficiency is closely related to size. Two independent hydrogen atoms occupy more space than a single hydrogen molecule. Two hydrogen molecules and a carbon atom occupy more space than a single methane molecule.

The process of going from big and bloated to small and compact releases energy. The more compact a configuration is, the less energy is left in it for further reactions. Inefficient, wide configurations, carry more potential energy than smaller denser ones.

Photons as Carriers of Energy

For a bouncing electron to go up one harmonic, it has to absorb a specific quantum of energy. To go down one harmonic, it has to release an energy quantum.

These energy quanta come in the form of photons.

In conventional physics, the photon absorbed or released is nothing more than a quantum of pure energy. It can therefore be created and destroyed whenever needed.

However, if photons are made of dielectric matter that can neither be created nor destroyed, the sudden appearance and disappearance of photons must be explained in some other way. There has to be a pool of photons available for the energy transfer.

Low energy photons must be everywhere present so that they can be kicked up in energy. However, they must not be so abundant that the energy transfer always happen immediately after an electron has been excited into a higher energy level.

The low energy photons have to be at a certain abundance corresponding to the typical time it takes for an excited electron to stay excited before returning to its lower energy level.

The process of excitement into a higher energy level, followed by the subsequent drop to a lower energy level will have to go as follows:

1. A random high energy photon hits a bouncing electron.
2. A quantum of energy is transferred from the photon to the electron.
3. The electron bounces at a higher harmonic.
4. A random low energy photon hits the excited electron.
5. A quantum of energy is transferred from the electron to the photon.
6. The electron bounces at a lower harmonic.

The time delay between step 2 and step 5 is determined by the availability of photons.

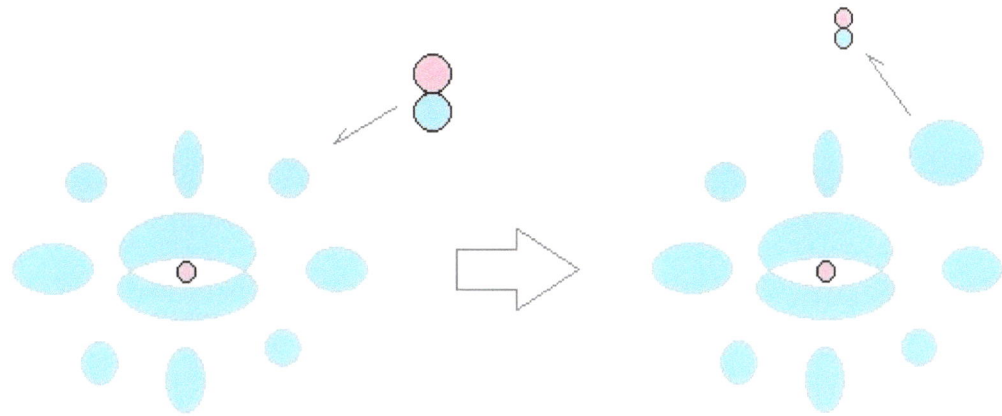

Electron of Neon being excited by an incoming high energy photon: step 1, 2 and 3

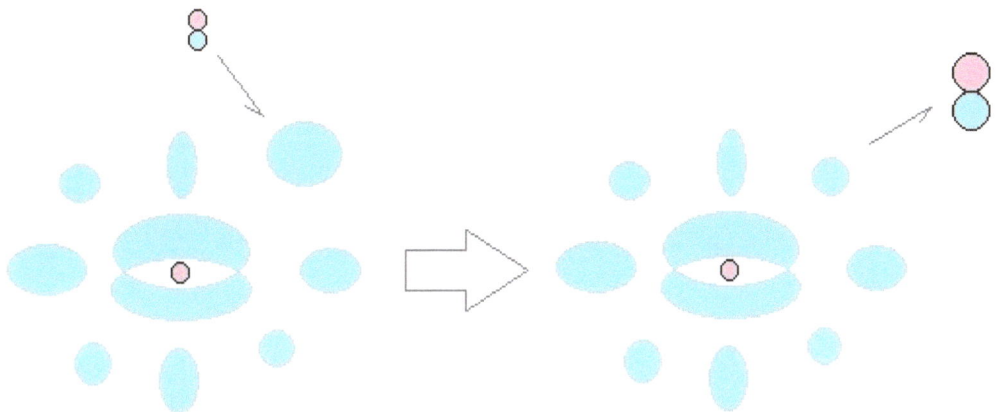

Exited electron of Neon kicking a random low energy photon up in energy: step 4, 5 and 6

For this to work, there must be a lot more photons around than is observed. They would have to be everywhere, and the vast majority of them would have to be in an undetectable state.

In short, we require an aether.

An Aether of Zero-Point Particles

There are many theories that invoke the aether in order to explain certain physical phenomena, and each theory has its own definition of what the aether is. Some theories require an aether in order to communicate energy in the form of waves. Other theories require it for other purposes.

The aether required for the physics laid out in this book is one in which there is an abundance of readily available low energy photons. I've chosen to call these particles zero-point photons to make the point that they have so little energy that they are undetectable. They may not be completely without energy. However, for practical purposes, they can be considered to have zero energy.

Zero-point photons fly about at the speed of light, just like any other photon.

Since space is known to be full of neutrinos, also flying about at the speed of light. The aether must be a mix of photons and neutrinos.

When it comes to neutrinos, the same logic applies as to photons. The vast majority of them are undetectable. We have zero-point neutrinos as well as zero-point photons. Collectively, we can refer to these as zero-point particles.

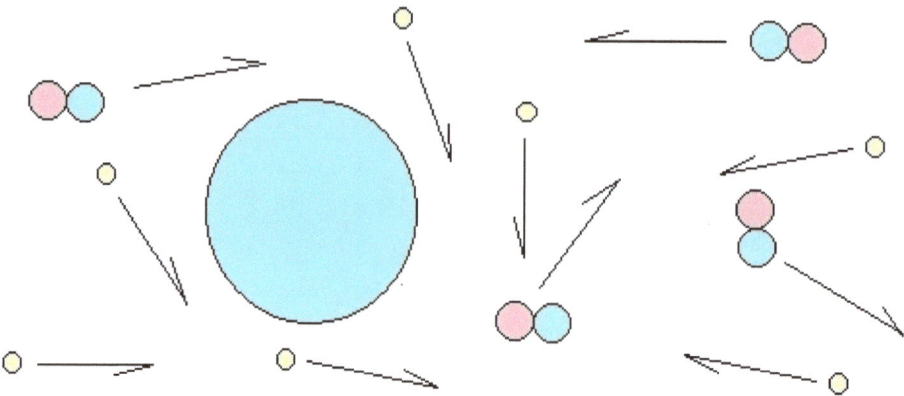

Electron surrounded by zero-point particles

These particles interfere with their detectable counterparts as well as ordinary matter.

Zero-point particles are very small. They have no trouble tunnelling through materials. They are therefore everywhere.

They are not directly detectable, but their effects are well known to us. In their bouncing about, zero-point particles form high and low pressure regions generally referred to as the electric, magnetic, and gravitational fields.

The Electric Field

The electric force can be explained entirely in terms of zero-point neutrinos. All that is required is to give the neutrino the ability to carry a footprint of whatever charged particle it has most recently interacted with.

When a neutrino hits a hook covered quantum, its hoops get drawn out. The neutrino gets a small charge which it carries with it back into space. A neutrino that hits a hoop covered quantum will return to space with its hooks drawn out.

The space surrounding a hook covered quantum is in this way filled with neutrinos with their hoops drawn out. The space surrounding a hoop covered quantum is full of neutrinos with hooks drawn out.

When charged neutrinos hit other charged neutrinos we get one of two types of collisions:

1. If the neutrinos carry different charges, we get an abrasive collision. The hooks and hoops latch briefly onto each other. They make a hard turn and vacate the field in between the two charged quanta. We get a low pressure in the aether.

2. If the neutrinos carry the same charge, we get a non-abrasive collision. There is no latching onto the other neutrino. There is no hard turn. The neutrinos stay in the field between the two charged quanta. This gives us a high pressure in the aether.

Low and high pressure regions are thus created in the aether by charged surfaces.

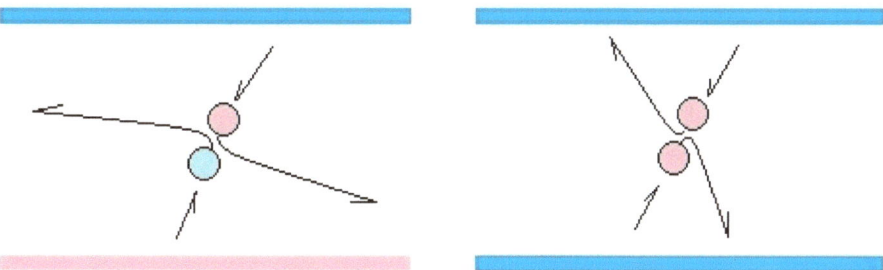

Abrasive and non-abrasive collisions of neutrinos between charged surfaces

This explains why opposite charges attract while same charge repel.

Static Charge and Neutral Bodies

When a neutral body comes in contact with a charged body, there is attraction. This may seem strange at first glance. However, it is relatively easy to explain in terms of charged quanta.

A neutral body is only neutral in so far as it has no net charge. Everything is made up of electrons and protons, which again are made up of positive and negative quanta.

A charged surface will pull opposite charges towards it. The distribution of charges in the neutral surface becomes distorted. Attracting charges rise to the surface while repelling charges withdraw into the material.

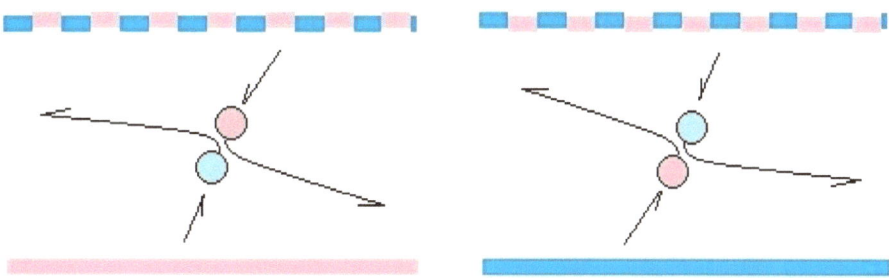

Neutral surfaces distorted and attracted by charged surfaces

With attracting charges closer to the charged surface than repelling charges, the net effect is attraction.

Each individual section of the neutral surface experiences either attraction or repulsion due to the charged surface. However, on average, the neutrinos inside the field will be of opposite charge due to the difference in distance between repelling and attracting sub-sections of the neutral surface. The majority of neutrinos collide abrasively, leave the field and produce low pressure.

Coulomb's Law

Coulomb's law states that the force between two point charges is proportional to the product of the two charges, divided by the square of the distance between them:

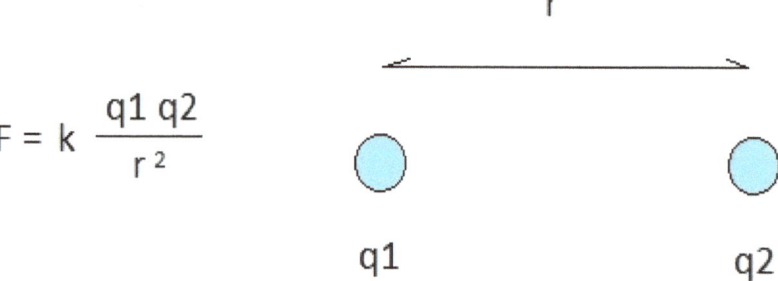

Coulomb's Law

This can be explained in terms of zero-point neutrinos, as follows:

- The density of charged neutrinos around a charged sphere falls off by the inverse square law. This can be derived directly from geometry. The surface area of a sphere increases with the square of its radius, thus reducing density by the inverse square law.
- The probability of a collision between two charged neutrinos, one from each charged sphere, depends on the number of charged neutrinos bouncing off of them. This in turn depends on the charge on the spheres themselves. Using basic probability theory, we get that the chance of a collision is directly related to the product of the two charges.
- The constant k is a measure of the availability of zero-point neutrinos.

From this we can explain Coulomb's law as follows:

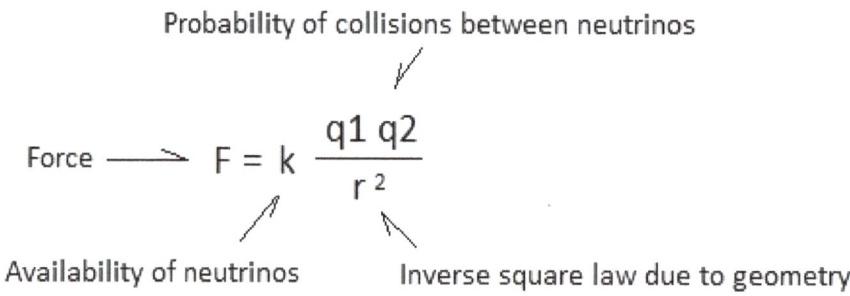

Coulomb's Law explained

It follows from this that Coulomb's law can be seen as supporting evidence for the position that the electric force is communicated by colliding particles.

Gravity

The electric force between an electron and a proton is due to their difference in charge. A proton carries a net charge of +1, while the electron carries a net charge of -1. The fact that there is a total of 2177 charged quanta making up the proton and a total of 3 charged quanta making up the electron does nothing to alter this. It is the net charge of the respective particles that matters.

Together, the proton and electron forms a neutral whole with a net charge of 0. However, this is not to say that there is no charge surrounding a neutral atom. There is a big difference between no net charge and no charge at all.

Neutral matter produce just as many charged neutrinos as charged matter. The only difference is that the charge adds up to exactly zero in the case of neutral matter, while charged matter produce an excess of either negatively or positively charged neutrinos.

The electric force depends on the net imbalance in charge between two bodies. When the number of positively and negatively charged neutrinos around a body average out to zero, there is no electric field.

However, as previously mentioned, there is a tiny difference in reactivity between positively and negatively charged quanta. This was illustrated with the analogy of Velcro, in which hooks react ever so lightly with other hooks while hoops don't react with other hoops. This in turn was used to explain why protons are larger than electrons.

Since positive quanta react lightly with each other, we get that a collision between two positively charged neutrinos will not be the completely perfect bounce that we get when two negatively charged neutrinos collide.

For two neutral bodies, we get that the following four types of collisions can happen with exact same probability. Note that all collisions except hooks on hooks produce one unit of pressure:

- Hooks meet hoops = 1 unit of low pressure
- Hoops meet hooks = 1 unit of low pressure
- Hoops meet hoops = 1 unit of high pressure
- Hooks meet hooks = 1-x unit of high pressure, where x is a tiny fraction of 1

The hooks on hooks collision produces a slightly imperfect collision, resulting in a less than perfect unit of high pressure. When we add up all the possible collisions, we get a tiny bit of low pressure.

With a sufficiently large number of collisions we get a weak attracting force.

It is this weak attracting force between neutral bodies that we refer to as gravity.

From this we see that gravity is a special case of the electric force. This in turn explains why the formula for Newton's universal law of gravity bears such a striking resemblance to Coulomb's law.

Coulomb's law is an expression for force based on net charge, while Newton's law is an expression for force based on total charge. Since inertial mass is directly related to the number of charged quanta making up protons and electrons, inertia is a perfect proxy for total charge.

$$F = k \frac{q_1 q_2}{r^2}$$ (Net charge)

Coulomb's Law

$$F = G \frac{M_1 M_2}{r^2}$$ (Total charge)

Newton's Universal Law of gravity

Coulomb's law compared to Newton's law

In conclusion, we can say that gravity is due to a tiny imbalance in the electric force.

Anti-Gravity

The imbalance in the electric force, which we call gravity, manifests itself as a low pressure area in the aether between bodies of dielectric matter. There is a tendency for neutrinos to leave the field between such bodies.

It follows from this that the regions away from the gravitational field must experience a high pressure corresponding to the low pressure. This high pressure is the opposite of gravity. It is anti-gravity.

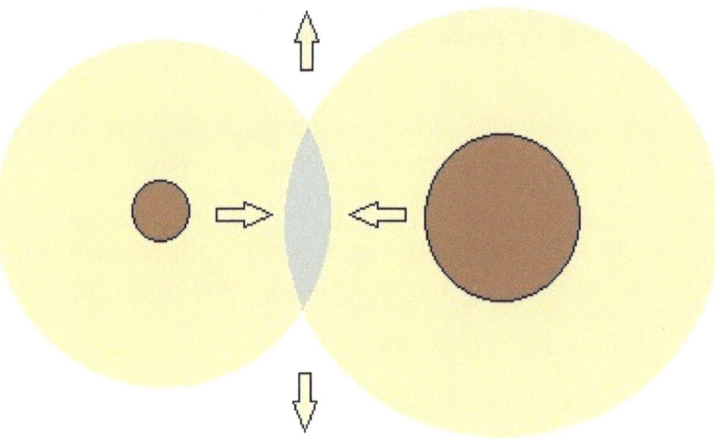

Gravity pulling bodies together, and anti-gravity dissipating into space

Since the space away from the gravitational field is much bigger than the field itself, the high pressure produced is dispersed to such a degree that it becomes impossible to detect in places like our solar system where astronomical bodies are fairly thinly distributed. However, in environments with a great number of astronomic bodies packed tightly together, anti-gravity may be detectable.

Gravity and Shielding

There is no way to shield ourselves from the effect of gravity. There is no material that we can stand on to prevent our planet from pulling on us.

This is because the force of gravity is a universally attracting force. The attracting force between two bodies may be consumed in the sense that their attraction only affects the two bodies in question. However, the effect is not lost. It daisy-chains out to other bodies.

Furthermore, the neutrinos involved are so small that the majority of them zip right through bodies without interacting.

When we stand on a slab of rock, the rock only "consumes" as much gravity as it "produces". In fact, a tiny bit of gravity is added to the total.

Every body of dielectric matter has a cloud of charged neutrinos around it. Where these clouds interact, we get a low pressure in the aether.

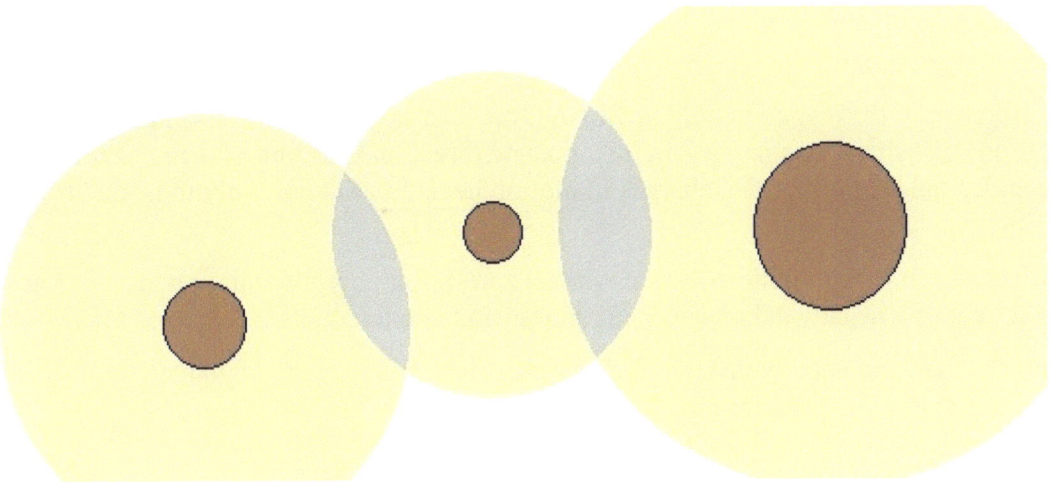

Gravity daisy-chaining between three bodies

Low pressure areas between bodies daisy-chain in such a way that the net effect can be calculated by treating each interaction individually before adding them all up to get the overall effect.

Electric Currents

An electric current can be defined as charge in motion. Normally when we think of electric currents, we think of electrons moving through a conducting wire. However, a positively charged gas, moving through space would also be a current. Any charged object or particle in motion constitutes an electric current.

For a current to be significant, it must have a large number of similarly charged particles, all moving at a similar velocity. The analogy is that of a river of charged particles.

For the purpose of definition, it has been decided that the direction of a current is that of a positive ion in motion. This means that when we have electrons moving through a wire in one direction, the current is by definition in the other direction.

The reason for this is the curious fact that a current always generates a magnetic field around it in such a way that if we hold our right hand thumb in the direction of the current, our fingers fold in the direction of the magnetic field. This is known as Ampère's right-hand grip rule.

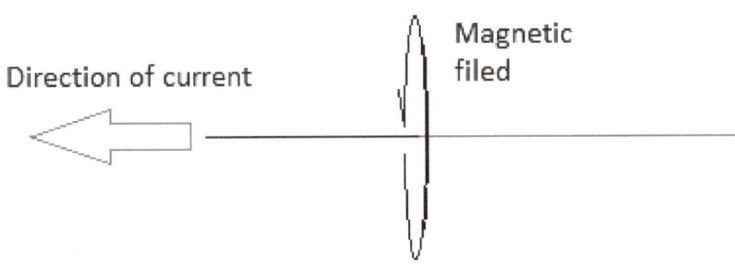

Ampère's right-hand grip rule

It does not matter if the current is due to positive ions moving from right to left, or electrons moving left to right. The magnetic field will always circle the current as if it was caused by positive ions moving from right to left.

This indifference of the magnetic field to whether the electric current is caused by electrons moving in one direction or positive ions moving in the opposite direction is at first glance puzzling. However, once we understand the effect that charges in motion have on zero-point photons, the mystery solves itself.

A positive charge moving in one direction will set zero-point photons spinning around it in a manner identical to that of a negative charge moving in the opposite direction.

The magnetic field accompanying all and every electric current can be explained entirely in terms of zero-point photons being polarized and set spinning.

The Two Orb Photon

When zero-point photons hit either protons or electrons, the hooks and hoops of the massive particles briefly latch onto the hooks and hoops of the photons. This cause the photons to change their spin and direction before continuing their way back into space.

For stationary protons and electrons, the change in direction and spin is completely random. There is no net effect. But for a charged particle in motion, there is a net effect. Zero-point photons will tend to spin in parallel with the moving particle. Spinning zero-point photons are what we observe as magnetic fields around electrical currents.

However, for this to happen in accordance to Ampère's right-hand grip rule, photons cannot be any random configuration of six charged quanta. Their structure must be of a very specific kind. It must be that of two counter-spinning orbs.

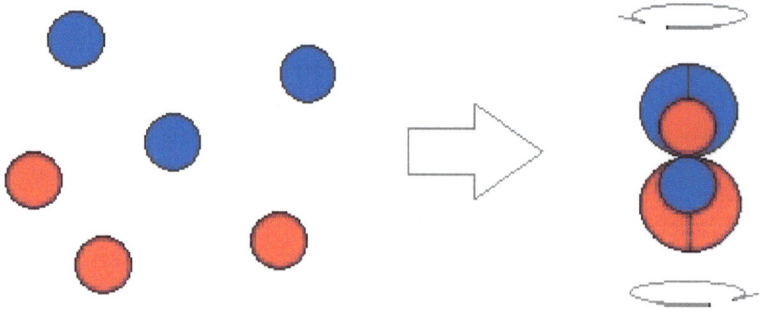

Two orb model of the photon

The six charged quanta making up the photon must be modelled as two orbs of opposite charge, one spinning one way and the other spinning the other way at the exact same rate.

If one orb latches onto a charged particle, thus changing its spin, the other orb changes its spin with an exact and opposite amount.

With this model of the photon, Ampère's right-hand grip rule becomes relatively easy to explain.

Ampère's Right-Hand Grip Rule

Key to understanding Ampère's right-hand grip rule in terms of zero-point photons is to imagine two ions moving through space. By comparing the effect of a positive ion moving from right to left, to the effect of a negative ion moving from left to right, we see that the two cases create identical magnetic fields.

A positive ion can be viewed as a hook covered ball. When it moves through space, it latches on to the hoop covered orbs of zero-point photons. A positive ion moving from right to left sets the negative orbs of the zero-point photons spinning counter-clockwise as viewed from the magnetic north.

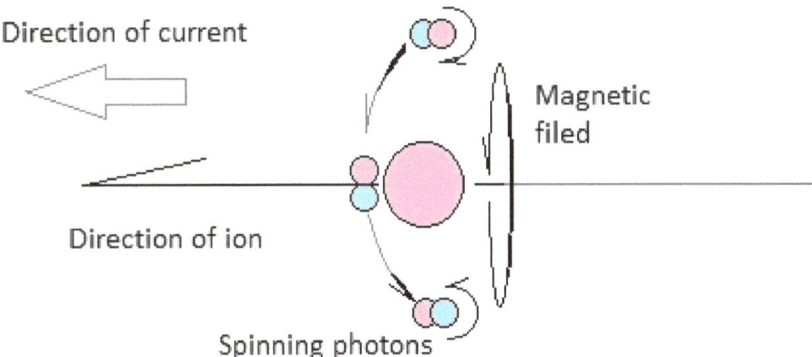

Positive ion producing magnetism in photons by setting their negative orbs spinning

Zero-point photons moving away from the photon after a bounce, are all polarized and spinning. A circular magnetic field radiates out from the moving ion.

Conversely, a negative ion moving from left to right, will set the positive orbs of zero-point photons spinning in the clockwise direction as viewed from the magnetic north. We get zero-point photons spinning with their negative orbs going counter-clockwise and their positive orbs going clockwise.

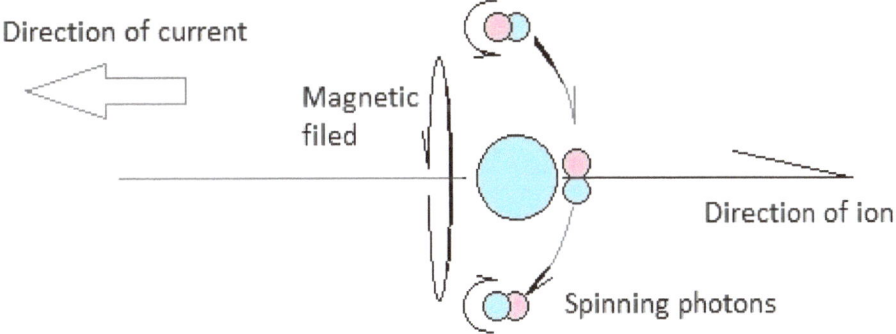

Negative ion producing magnetism in photons by setting their positive orbs spinning

The photons spin the same way in both cases. The magnetic field developed around the two ions are identical.

The two orb model of the photon behaves precisely the way it has to in order to model real world magnetism around charges in motion.

Permanent Magnets

A permanent magnet is a piece of metal, usually iron, which has a permanent magnetic field associated with it.

The way such materials can be explained in terms of zero-point photons, is that the atoms making up a permanent magnet are arranged in such a way that their electrons hook onto one side of nearby zero-point photons more readily than the other side.

The more coordinated and vigorous the atoms are in their lopsided effect on zero-point photons, the stronger the magnet.

Bouncing about inside a magnet, zero-point photons become polarized. They are given direction and spin. The photons are led into paths going in a north-south direction.

The result is a stream of polarized zero-point photons exiting the magnet from both poles in equal measure.

Magnet inducing spin into photons streaming out of the south and north poles

Correspondingly, there must be entry points for photons at the poles. Otherwise, there would be a permanent photon high-pressure at the poles and a corresponding low-pressure at the sides. This would violate the laws of thermodynamics, and is obviously not happening.

When outgoing polarized photons meet incoming photons, they brush into them, sharing some of their spin. This polarizes the incoming photons as they head towards the magnet. Even before they enter the magnet, they have a certain degree of polarization.

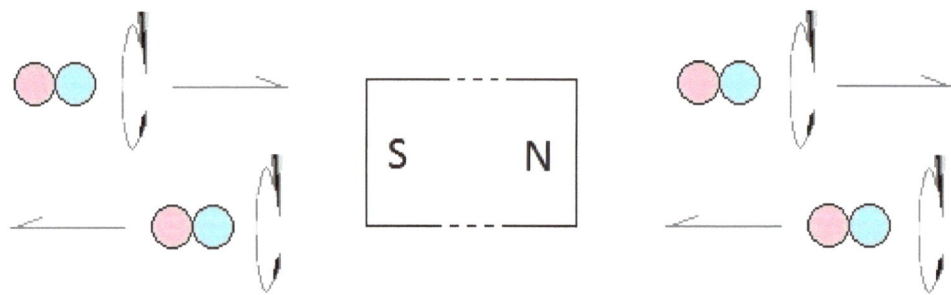

Photons entering and leaving both ends of a magnet

Note that spin is transferred between orbs of opposite charge. Orbs with identical charge do not react with each other since they cannot latch onto each other. However, negative orbs communicate their spin to positive orbs and visa versa. This allows for spin to be maintained and shared.

The sharing of spin from outgoing to incoming photons produces a pattern in which highly polarized outgoing photons are surrounded by progressively less polarized photons. Between each highly polarized outgoing photon, there is a valley, so to speak, of less polarized photons.

To see a manifestation of this pattern, all we have to do is to put a ferro-fluid on top of a magnet.

Ferro-Fluids

When a ferro-fluid is placed on top of a magnet, it morphs into sharp peaks surrounded by shallow valleys. Highly polarized outgoing photons are producing the peaks, while less polarized incoming photons are producing the valleys.

Ferro-fluid

By Steve Jurvetson - http://www.flickr.com/photos/jurvetson/136481113/, CC BY 2.0, https://commons.wikimedia.org/w/index.php?curid=906519

The Faraday Effect

The fact that photons do not have to directly hit a magnet in order to become polarized was discovered by Michael Faraday back in 1845. By passing visual light through a magnetic field, he observed polarization. This is what we call the Faraday Effect. It can be readily reproduced, and proves that there is a direct relationship between magnetism and light.

The way this is interpreted in the physics laid out in this book is that highly polarized zero-point photons exiting a magnet will rub against visible photons, thereby transferring some of their spin and polarization to the visible light.

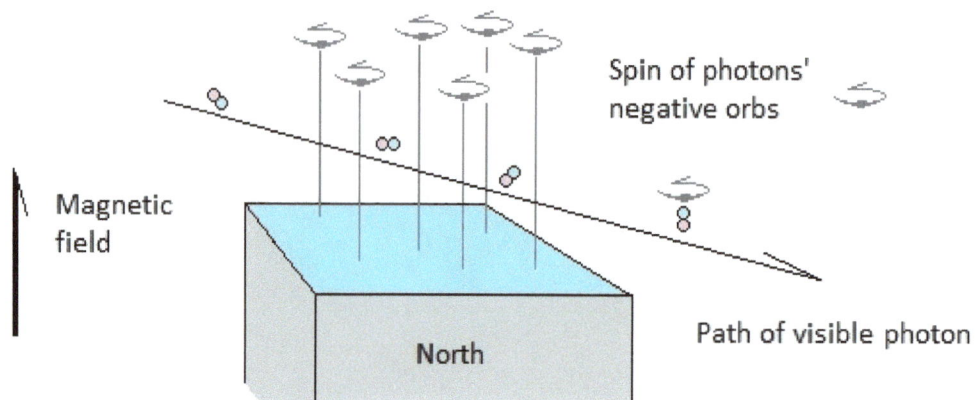

Visible light polarized by a magnetic field

The Faraday Effect can in other words be seen as supporting evidence for the position that magnetism and polarized light is one and the same thing.

When photons are polarized so that their positive and negative orbs spin along the same axis and with their positive and negative orbs facing the same way, we have magnetism.

The Magnetic Field

Magnetic fields fan out to the side. Outgoing photons allow incoming photons to come in between them. The outgoing photons yield to the incoming ones.

If we place a bar magnet under a sheet of paper, and sprinkle iron filings on top of it, we can see how this fanning out continues in all directions so that we get a pattern that connects the north pole to the south pole.

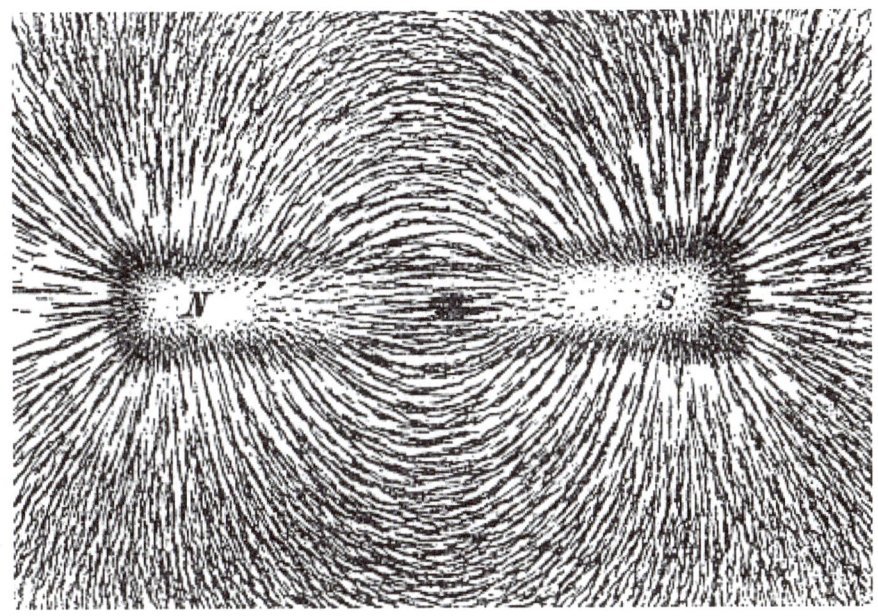

Magnetic field lines illustrated by iron filings on paper above a magnet.

By Newton Henry Black - Newton Henry Black, Harvey N. Davis (1913) Practical Physics, The MacMillan Co., USA, p. 242, fig. 200, Public Domain, https://commons.wikimedia.org/w/index.php?curid=73846

Polarized photons stream out from the north and the south pole of magnets in equal measure. However, there is no overall flow. All that is happening is that the polarized photons arrange themselves in the most efficient manner possible. Magnets polarize zero-point photons which in turn polarize all photons in the entire space around the magnet.

If two magnets are placed so that their north poles or south poles face each other, polarized photons from the magnets will meet head on in a non-abrasive collision. Since the colliding orbs are of the same charge, there is no latching onto each other. There are no hard turns, so the photons will tend to stay in the field. The result is a high pressure in the aether between the magnets. We have a repelling force.

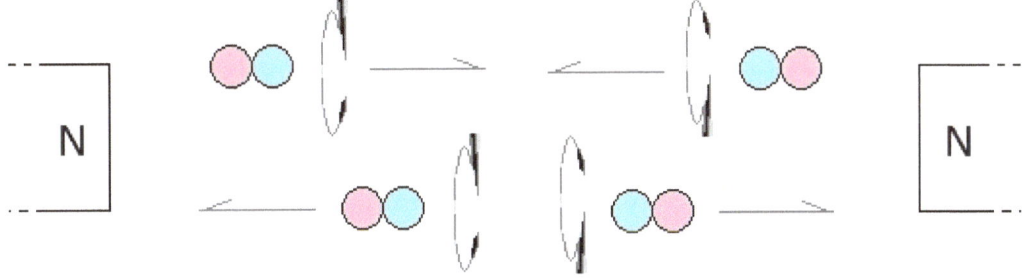

Non-abrasive collisions between photons produce high pressure

Conversely, if a north pole is facing a south pole, polarized photons collide with hooks against hoops. The collisions are abrasive. The photons latch onto each other. They make a hard turn and exit the field. This results in a low pressure area in the aether between the magnets. We have an attracting force.

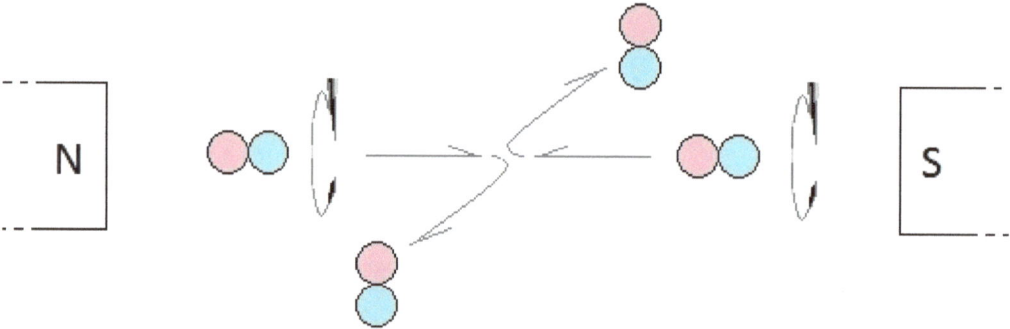

Abrasive collisions between photons produce low pressure

This is identical to how neutrinos produce high pressure and low pressure areas in the aether through collisions. The magnetic force is communicated by photons in the exact same way that the electric force is communicated by neutrinos.

Magnetic Materials

Most materials react in some way to magnetic fields. However, most of these react so weakly that it can only be noticed under very controlled conditions.

The most common reaction to a magnetic field is a minuscule repelling force. These are the so called diamagnetic materials.

Then there are the paramagnetic materials that react with a minuscule attracting force.

Finally, we have the ferromagnetic materials such as iron that react with a strong attracting force.

This can all be explained with polarized zero-point photons being reflected by the materials involved.

In the case of diamagnetic materials, we have reflection in which the polarized photons are flipped around. This produces a repelling force.

Diamagnetic materials: polarized photons are flipped on reflection

In the case of paramagnetic and ferromagnetic materials, we have reflection in which the polarized photons are bounced back directly. There is no flipping around of polarity, and we get attraction as a result.

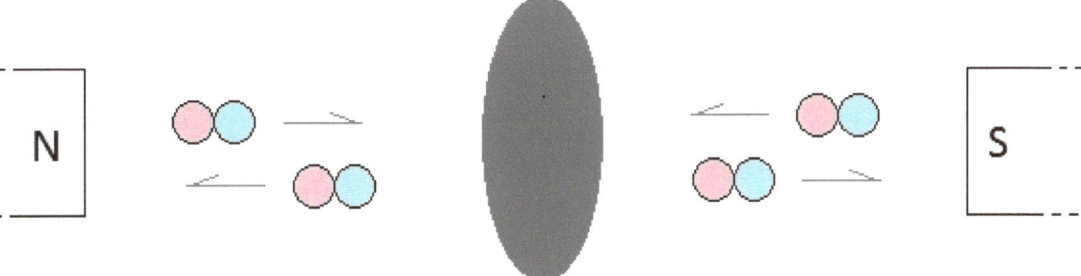

Paramagnetic and ferromagnetic materials: polarized photons are not flipped on reflection

In the case of diamagnetic and paramagnetic materials, the effect is weak. There is a lot of scatter and little reflection. However, in the case of ferromagnetic materials, reflection is strong.

Unlike visible light, which reflects off of the outer surfaces of materials, zero-point photons reflect mostly off of the atoms inside the material.

Zero-point photons are very small compared to visible light. They typically tunnel into materials before interacting with it. There is no way to polish an outer surface to improve the magnetic properties of a material.

Making a Permanent Magnet

In addition to being highly reflective of zero-point photons, ferromagnetic materials can be induced with a permanent magnetic field. Their atoms can be such arranged that zero-point photons get polarized when inside the material. This is done by exposing a ferromagnetic material to a strong magnetic filed.

From our knowledge of how Ampère's right-hand rule works, we can see that such a field can be produced with electricity. If we wind a coil of copper wire around a piece of ferromagnetic material we get the required magnetic field by applying a direct current to the copper wire.

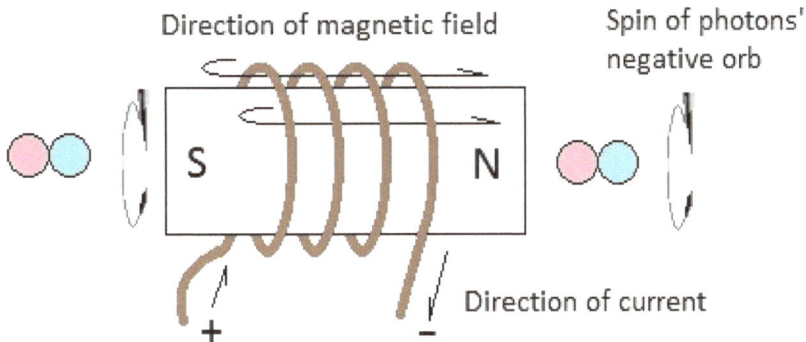

Induction of magnetism into a ferromagnetic rod

Highly polarized zero-point photons spin their way through the ferromagnetic material, pushing the atoms into place, thus inducing permanent magnetism into the material. Only ferromagnetic materials are receptive to this kind of treatment.

Induction of Currents into Wires

To illustrate how magnets can be used to induce electric current into a copper wire, we can start by imagining a copper wire connected to an ammeter to measure current.

If we place this wire at rest on top of the north or south pole of a magnet, nothing happens. There is no measurable effect if nothing moves.

However, if either the magnet or the wire is moved in such a way that the copper wire cuts into the stream of polarized photons coming out of the magnet, then we will register a current.

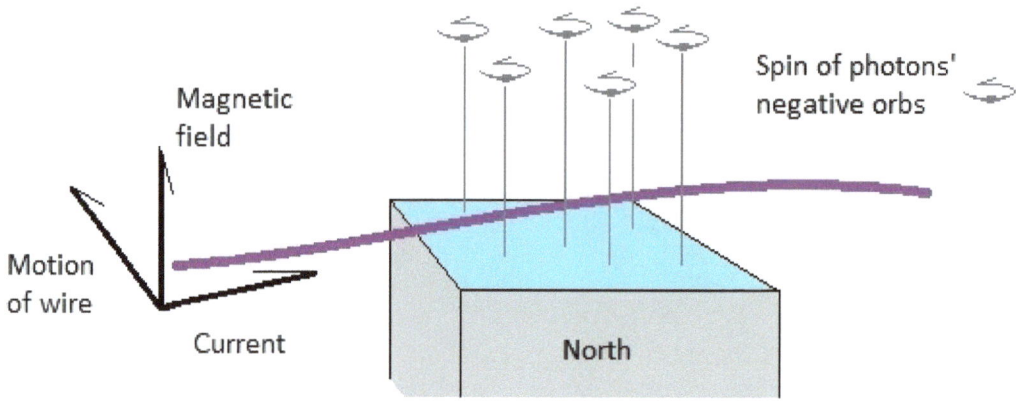

Induction of current into a wire

This can be explained entirely in terms of polarized zero-point photons interacting with electrons in the wire.

Moving the wire into the magnetic field, as illustrated above, makes the positive orbs of the spinning photons hook into the side of electrons in the wire. The electrons move to the left. A current towards the right is thus induced into the wire.

The current moves in the direction dictated by the spin of the negative orbs.

Move the wire in the opposite direction and the spinning photons hook into the electrons from the opposite side. The current moves the other way.

Flipping the magnet around so that it points down will likewise send the current in the opposite direction.

The relationship between motion of a wire, magnetic field, and current induced is always the same.

This is the right hand rule of electromagnetism, which states that an open right hand with thumb perpendicular to fingers can be used to determine the direction of induced current.

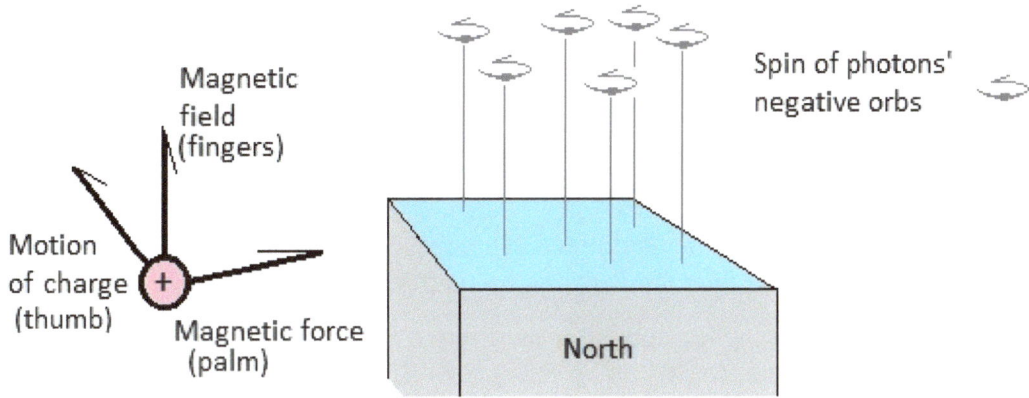

Right hand rule of electromagnetism

Let the thumb be the direction of motion of a positive charge, let the fingers be the direction of the magnetic field lines. Then the palm of the hand becomes the direction of the induced force.

In the case of electric induction, the wire is the charge in motion and the induced force is the current.

In the case of an electric motor, the current is the charge in motion and the induced force is the generated mechanical motion.

Motors and Generators

Now that we have seen how electric currents produce magnetism, and magnetism produce electricity, we can make electric motors and generators.

Kinetic energy of a waterfall or other energy source can be turned into electricity by a generator. The electricity can be transported by a copper wire to a distant location. An electric motor can in turn be set moving.

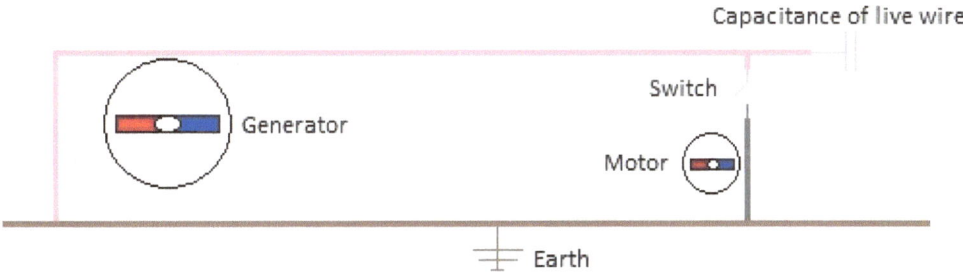

Circuit diagram of generator and motor

In the above circuit diagram, we have a generator consisting of a spinning magnet. This induces an alternating current into the live wire by sending electrons moving one way when the north pole of the magnet sweeps past the wire, and sending electrons moving the opposite direction when the south pole sweeps past the wire.

As long as the switch to the electric motor is open, nothing much happens. The electrons in the wire are pushed back and forth a bit, but there is no load on the generator. Very little energy is consumed.

However, once the switch is closed, electricity starts flowing freely past the magnet inside the motor.

The strong magnetic field induced by the free flow of electrons set the magnet in the motor spinning.

The physical gaps between magnet and wire for both generator and motor are bridged by strong magnetic fields. The load on the motor is thus passed on to the generator.

Radio Transmission

When an alternating current is sent up and down an antenna, an alternating magnetic field is created around the antenna in accordance with Ampère's right hand grip rule.

When the current is going up the antenna, photons are sent off spinning one way. When the current is going down the antenna, photons are sent off spinning the other way.

The polarized photons rush off in all directions at the speed of light.

When some of these photons hit a receiving antenna at a distance, they induce electricity into it. The received signal matches the transmitted signal exactly. It can therefore be used to reproduce the transmitted message.

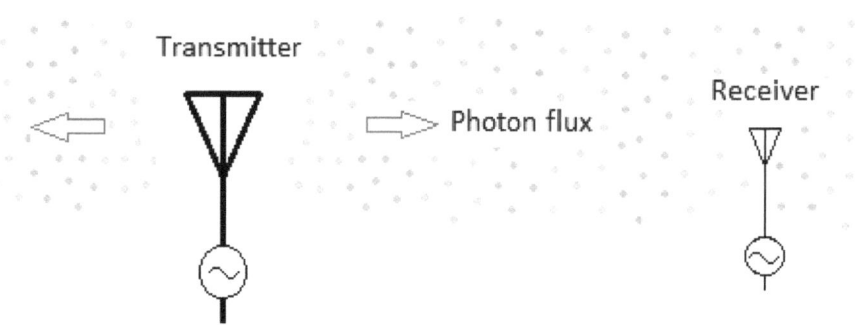

Radio transmission

This is how radio transmission works.

It should be noted that this can only work if the photons crossing the space between the transmitter and the receiver carry some energy with them. This is because electrons have inertia. It takes energy together with spin to set them into motion.

Since zero-point photons carry no energy, we know that the photons used in radio transmission cannot be zero-point photons. Radio transmitters do not only set zero-point photons spinning, they pump their energy up too.

The energy required is very small. Radio-wave photons are the least energetic of all detectable photons. They are just one step up in energy from zero-point photons.

Note that while electric generators and motors work by creating strong magnetic fields to bridge the gaps between components, radio transmitters work by giving energy to polarized photons. The large distance between transmitter and receiver makes it necessary to send energy together with the photons involved.

The tiny gaps between stators and rotors in generators and motors are bridged by magnetic fields. No added energy is required for this to work since magnetic fields are nothing but low and high pressure regions in the aether.

The Faraday Cage

A Faraday cage is a metal casing used to protect whatever is inside it from electric and magnetic forces. A well designed Faraday cage can protect a person from lightning as well as external radiation.

It is not a perfect insulator. Radio-waves and some magnetic fields can penetrate a Faraday cage with various degrees of attenuation. High energy photons such as x-rays are largely unaffected by a Faraday cage.

In the case of electricity, Faraday cages are easy to explain. The metal of the cage leads the electricity that strikes the cage to ground, without going through the interior of the cage. The cage is a sufficiently good conductor to take care of the electricity, and lead it safely to ground.

In the case of low energy radiation, the cage acts like a receiving antenna. It transforms the energy of radio-waves to electricity, which in turn is led to ground.

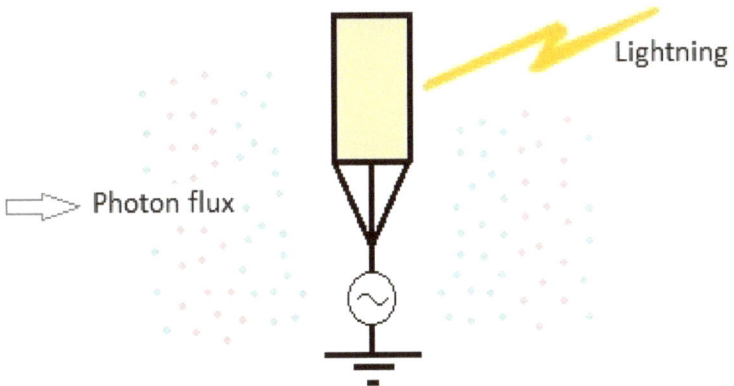

Faraday cage protecting its interior by leading electricity to ground

Such a cage will only protect against x-rays to the extent that it is built from a material sufficiently impregnable to high energy radiation.

Size of Photons

If neutrinos make up the electric force that keeps molecules from flying apart, they must be very small and also very abundant. They must be everywhere zipping back and forth, communicating the electric force on their way from one place to another.

Similarly, zero-point photons must be both abundant and small in order to be available inside materials to be kicked up in energy, and to communicate the magnetic force in the presence of moving charges.

These particles, which constitute the aether, must be the smallest of all particles. Being undetectable, except through their manifestation as force, they must also be without energy, or with so little energy that they cannot be detected.

One up in energy from zero-point photons, we have radio-wave photons. These particles are known to pass through most materials. Cell phones work inside buildings, cars and elevators. Radio-wave photons are almost as good at passing through materials as zero-point particles.

Radio-wave photons appear to be only a little bigger than zero-point photons.

Then there is visible light. These photons have more energy than radio-wave photons. The fact that they have problems going through most materials suggests to us that they are larger.

The most energetic photons are x-rays and gamma-rays. These are able to go through many materials. However, they are destructive. They go through walls like bullets through a net. They may not hit anything, in which case no damage is done, or they hit something and cause damage.

Unlike radio-waves, x-rays rarely change their direction on travelling through a material. They get through it in a straight line or they are stopped by hitting into something on their way. There is little scatter.

This is why x-rays are used to make pictures of bones and the like inside our bodies. They produce nice sharp images, shaded according to how many of the x-ray photons were stopped on their way.

However, their destructive nature prevent us from taking such pictures frequently. They can cause cancer and radiation sickness.

Even more dangerous are gamma-rays. They are the most energetic of all photons. So energetic that they sometimes pop, in which case they make the transition from being photons to being an electron-positron pair.

It is as if gamma-ray photons are so large that they cannot be made larger. Any collision that adds energy to a gamma-ray photon causes it to pop.

From this, we can conclude that the energy of photons are directly related to their size. The more energetic a photon is, the bigger it is.

As we will see, this can be used to explain both refraction and diffraction in optics.

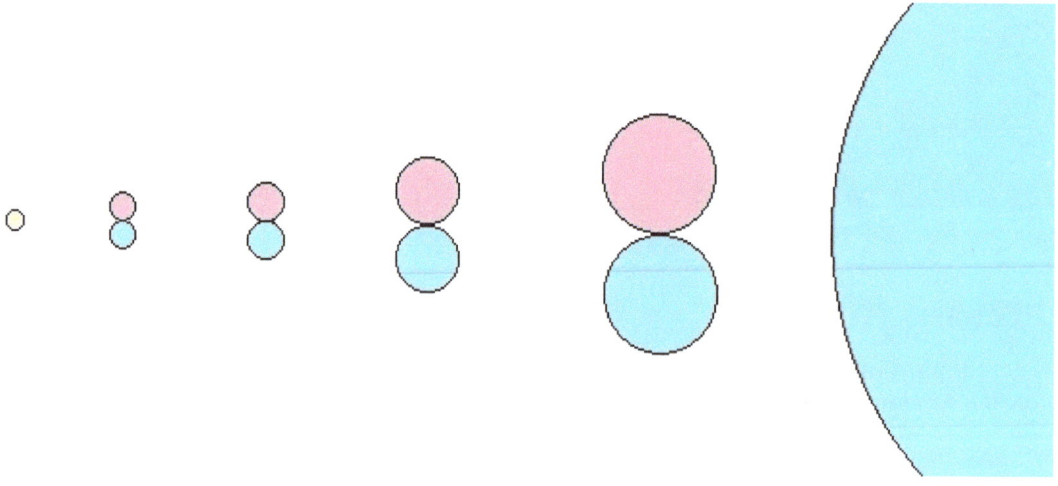

Neutrino, zero-point photon, radio-wave photon, visible photon, gamma-ray photon and electron

Light Travelling Through a Transparent Medium

Transparent materials such as glass and water have the ability to let light through as if they were made of nothing. Yet some transparent materials, such as glass, are very dense. They are full of atoms.

For light to pass through such materials without scatter, every photon has to meander through the material in a fashion identical to every other photon.

A way to envision this is to think of photons as slalom skiers, and the glass as a slope full of evenly spaced poles in all directions, with the poles being atoms.

When a photon enters such a material, it starts with a half roll past the first atom. Then it continues with a full roll past every subsequent atom until it makes a final half roll past the last atom before exiting.

The photon may make a first half roll to the left or the right. It does not matter. However, the next roll has to be in the other direction, and the next roll after that has to be opposite to the previous, and so on all the way through the material.

For photons entering the material at an angle, the first half roll will either be larger or smaller than average, depending on the angle of entry and which side of the first atom they enter. However, this is perfectly balanced on exit with a corresponding deviation from the average.

This will result in all photons leaving the material in the exact same direction that they entered it, provided the first row of atoms are parallel to the last row of atoms.

Since photons travel at the exact same speed regardless of their size. They always travel at the speed of light. However, the length of the path travelled by a small photon and a big photon will not be identical.

Small photons roll past atoms with their geometrical centre closer to the atom than the bigger photons, so even when large photons and small photons take the exact same path through a transparent medium, the smaller ones end up travelling a shorter distance.

Send a red photon and a blue photon through a piece of glass at the exact same time, and the red one ends up exiting the glass ahead of the blue one. The red one has less energy than the blue one. It is smaller, and is therefore rolling past the atoms in the glass at a shorter distance from the atoms' centre than the blue one.

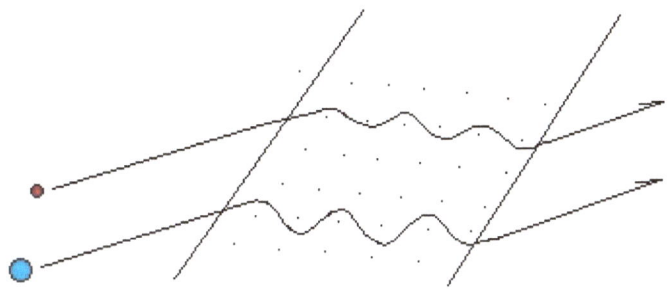

Red and blue photons racing through a piece of glass

This explains why blue light takes more time to travel through transparent media than red light.

It also explains why blue light refracts more through a prism than red light. It explains why a mix of various size photons, known to us as white light, get split into all the colours of the rainbow, with blue light always at the most acute angle from the prism, and red light at the least acute angle.

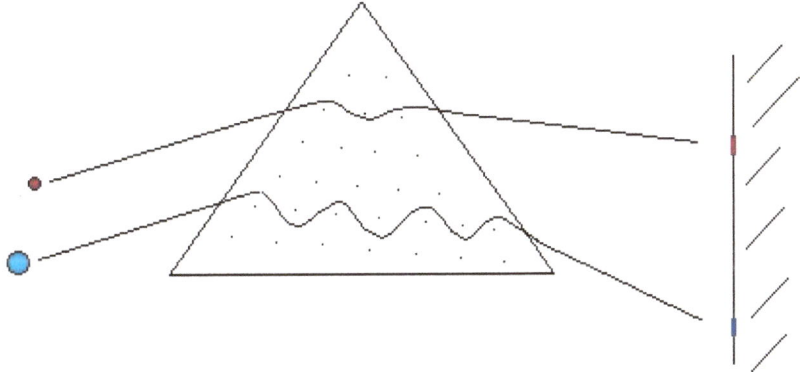

Photons hitting a wall after travelling through a prism

Being larger than red photons, blue photons take more time rolling past the first atom. This makes the initial half roll more acute for blue photons than red photons. It also makes the full rolls and the final half roll more acute.

The initial and final half roll of photons are precisely defined by the photons' size compared to the atoms in the medium. The bigger the photons, the more acute are their half rolls into and out of the prism.

Note that the photons do not divert from each other in their overall direction on entering a medium. Photons of different colours race through the medium in parallel.

It is not until the final half roll that diffraction happens. If the final half roll is back into the original direction, as is the case with plain glass sheets, the difference in original half roll entering the glass is cancelled out by the difference in half roll on exiting the glass.

However, if the final half roll is to the same side as the original half roll on entering the glass, as is the case in a prism, the original angle does not cancel out. It gets added to, and there is diffraction.

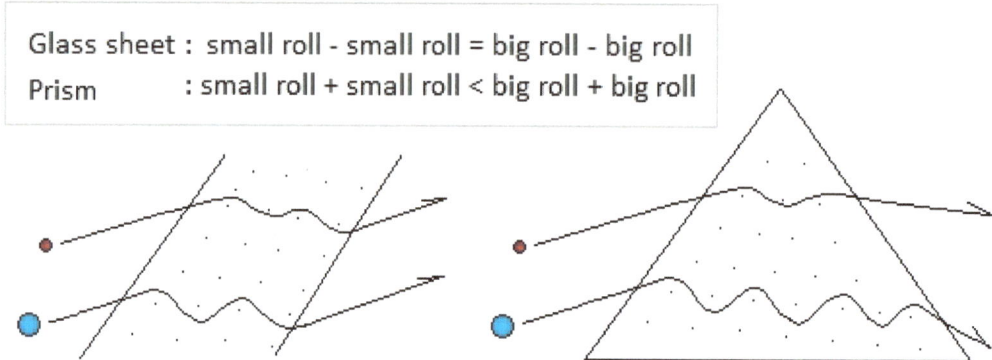

Diffraction happens as photons exit the media, and only when half rolls do not cancel on exit

This is why white light remains sharp and focused even through the thickest of glass sheets, while the smallest of prisms split white light just as well as big ones.

Note also that this has nothing to do with wavelength. All that matters is the size of the photons.

Snell's Law

Snell's law is a formula used to describe the relationship between the angles of incidence and refraction, when referring to waves passing through a boundary between two different isotropic media, such as water, glass, or air.

This formula is easy to establish in a wave lab with waves passing between deep and shallow water.

Since white light splits into different colours when passed through a prism, Snell's law is often used as if light is a wave phenomenon.

However, light does not in fact act like waves. White light does not split into different colours when entering a prism, as Snell's law would demand if white light was made up of many different waves. White light splits only at the exit.

If it was any other way, we would have noticed this long time ago. Plane glass sheets would have red light come out in a different place than blue light. Thick glass sheets, such as those used in large aquariums, would have multi-coloured fishes look like big smudges when viewed at an angle.

The fact that plane glass sheets do not smudge images, even when viewed at an angle, proves that light does not split on entry into glass. It splits only on exit, and only if the angle of exit is different from the angle of entry.

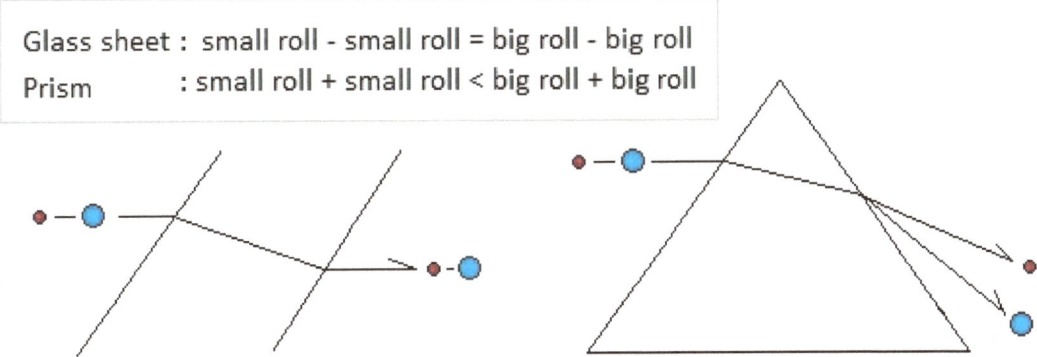

Red and blue light travelling through a plane glass sheet and a prism

Snell's law can therefore be used as supporting evidence for the position that light is not a wave phenomenon, but particles. The slalom analogy is better at explaining what's going on than the wave analogy.

The Double Slit Experiment

The double slit experiment has been used as "proof" that not just photons, but all things have wave-like properties. The larger an object is, the smaller is its frequency. Many things are so large that their wavelength cannot be detected. However, according to this wave-theory of matter, all things have wavelength.

Many double slit experiments have been performed, and they all seem to verify this theory. Red photons have the longest wavelength. Blue photons have shorter wavelengths. Electrons have shorter wavelengths still. Atomic nuclei have wavelengths too, and even molecules have been measured to have wavelengths, extremely small, but detectable.

However, all that this proves is that there is interference taking place when particles are passed through narrow slits, and that this interference is related to the size of the particles involved.

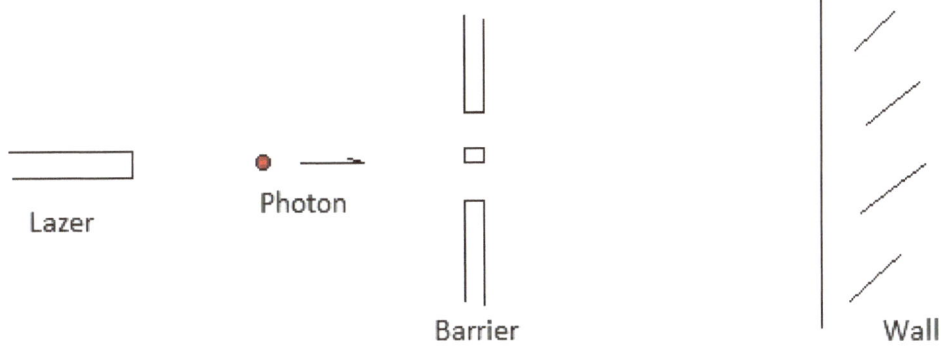

Double slit experiment set-up

What baffles people is the fact that an interference pattern appears at the detector/wall at the far end of the double slit experiment set-up even if only one particle is let through the slits in the barrier at a time.

But this is only a mystery if there's no aether. As soon as we include an aether of zero-point particles, the mystery of the observed interference pattern goes away.

Since zero-point particles come in two types, namely neutrinos and zero-point photons, they resonate with each other. These particles form a standing wave in which certain regions of space are more likely to contain a zero-point photon than other regions.

When a relatively large particle is sent through space, it bobs along on the standing wave. This creates a disturbance in the standing wave that propagates together with the particle.

This disturbance passes through the two slits like a wave in a lake. The particle moves like a vessel through these waves.

The larger the particle, the less it is affected by the waves, and the tighter is the pattern observed at the detector. Blue light produces a tighter pattern than red light. Electrons produce tighter patterns than photons. Atoms produce tighter patterns than electrons, and molecules produce the tightest of all patterns.

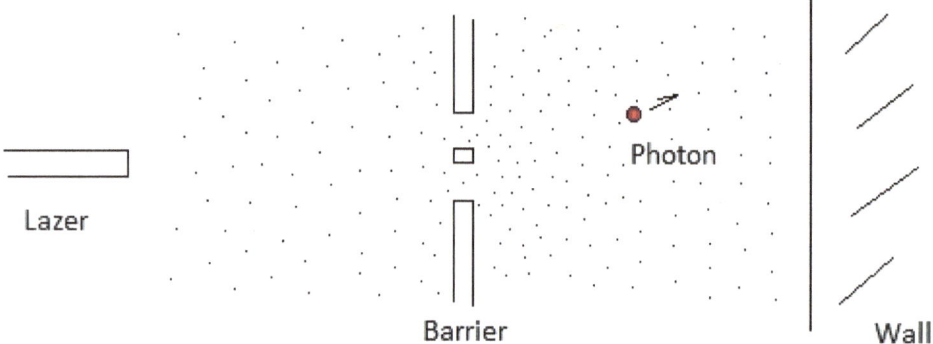

Photon bobbing along on a disturbed standing wave of zero-point particles

What is detected at the receiver is not the wavelengths of particles, but the relative size of detectable particles compared to zero-point particles.

Pilot Wave Theory

The idea that there is a pilot wave accompanying every particle is not new. Louis de Broglie made this suggestion back in 1929. He did this in order to get around the bizarre idea that particles can interfere with themselves by going through two openings at the same time.

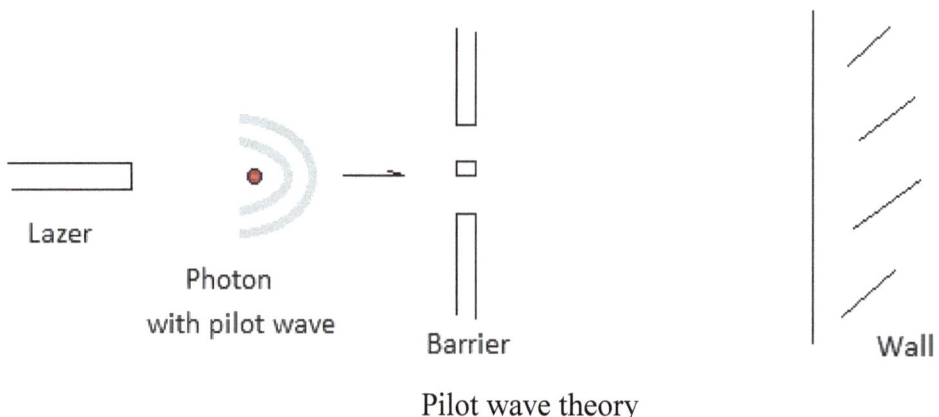

Pilot wave theory

The main difference between de Broglie's pilot wave theory and the mechanism outlined in this book is that de Broglie never gave an explanation as to what the pilot wave might consist of. Instead of an aether, he invoked a mysterious hidden variable.

Neutrinos, Photons and Space

In a physics that invokes an aether, there is no such thing as empty space. The aether is by definition something that is present everywhere. In the case of the physics laid out in this book, the aether is a mix of neutrinos and zero-point photons.

Space can therefore be considered a transparent medium. Photons travelling through "empty space" are in fact moving through a sea of very low energy particles.

This means that the same rules that apply to transparent media apply to space.

Small photons travel in a straighter line than big photons as they roll past zero-point particles. Small photons will therefore always win in a race with bigger photons even though these particles all travel at the same speed.

The shorter distance traversed by the smaller particle is what makes it arrive quicker at the destination.

It follows from this that the neutrinos, being the smallest of all particles, travel the fastest. In a race between a neutrino and a photon, the neutrino will come out the winner.

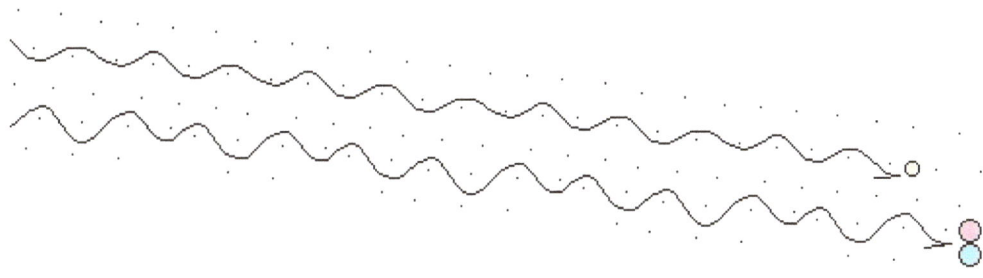

Neutrino in the process of overtaking a photon

However, the effect is likely to be minuscule. Zero-point particles are extremely small. Enormous distances may be required to come anything close to the effect of a thin sheet of glass.

Neutrinos, Photons and Gravity

We have already established that gravity is due to a small imbalance in the electrical force, and that this force is communicated by neutrinos.

A consequence of this is that neutrinos are not affected by gravity. They carry gravity. They are not themselves affected by it.

Gravity is a force that only affects dielectric matter. Neutrinos are not dielectric. When they carry charge, it is only one charge at the time. They never carry two charges at the same time.

Photons, on the other hand are dielectric. Just like electrons and protons, photons consist of a combination of charged quanta. Photons are therefore affected by gravity.

Every body of dielectric matter has charged neutrinos surrounding it. These neutrinos form low pressure areas in the aether on collision with each other. This is also the case for photons.

Since neutrinos travel slightly faster than photons due to their smaller size, a cloud of charged neutrinos extends in all directions from the photon, including the front of it. This cloud is the pilot wave mentioned earlier.

When a photon, surrounded by its pilot wave comes into contact with a charged cloud emanating from a massive body, the photon is affected by the resulting low pressure area in the aether just like any other bit of dielectric matter.

However, photons cannot slow down or speed up. They have to travel at the speed of light. A photon moving away from a massive body must therefore shed energy in some other way than slowing down. It has to go down in energy by becoming smaller.

Conversely, a photon approaching a massive body must absorb energy by becoming larger.

This is what we observe as gravitational red-shift. Light escaping massive bodies becomes redder. Light approaching massive bodies becomes bluer.

Light moving past a massive body bends off in the direction of the massive body, not due to a curvature of space, but due to normal gravitational behaviour of dielectric matter.

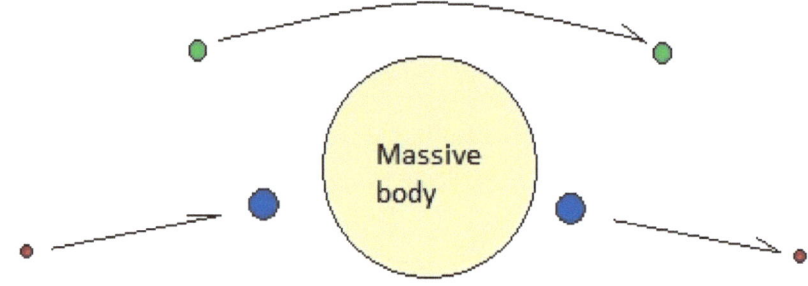

Photons passing by, moving towards and moving away from a massive body

A consequence of this is that the aether close to massive bodies are richer in zero-point photons than the aether farther away.

Gravity pulls all photons, including zero-point photons towards massive bodies. Neutrinos and zero-point photons are therefore not evenly distributed throughout space. Space far away from massive bodies have more neutrinos and fewer photons than space close to massive bodies.

Red-Shift of Light

The fact that photons cannot speed up or slow down explains why light escaping massive bodies turn red. It also explains why bodies moving away from us appear redder than bodies moving towards us.

The way to think about this is to first consider ordinary matter, and then apply what we know to be true for such bodies to light.

In cases where ordinary matter would normally absorb energy by speeding up, light will turn bluer. In cases where ordinary matter would normally loose energy and slow down, light will turn redder.

Since a ball tossed towards us from an oncoming object will move faster than a ball tossed towards us from a receding body, we know that photons reflected by a body approaching us will be bluer than photons reflected by a receding body.

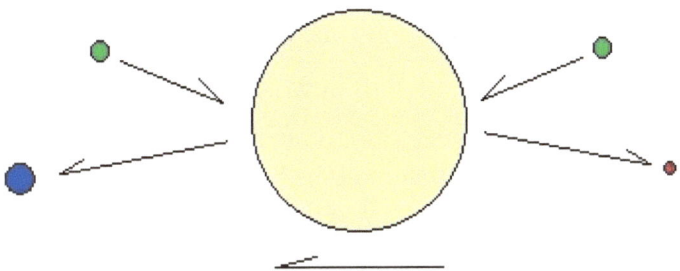

Green photons reflected by body moving from right to left

Keep in mind that all photons can be considered to be reflected, also those that are excited by a lamp or radio transmitter. Photons originate in the aether, and are therefore reflected by bodies, regardless of whether they are exited or not.

It should also be noted that it does not matter whether it is the observer or the observed object that moves.

If the observer travels towards an object, the photons will be registered as bluer than if the observer travels away from the object.

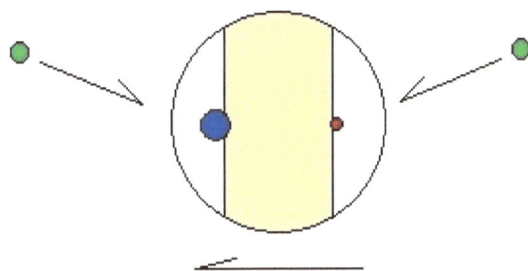

Green photons registered by detector moving from right to left

This is because the observer adds energy to the interaction with the approaching photon when moving towards it. Conversely, the observer subtracts energy from the interaction when moving away from a photon.

Red-shift and blue-shift of light is therefore entirely about relative motion. It does not matter what energy level the photon has while traversing the space between the observer and the observed. All that matters is relative motion. If the observer and the observed travel at identical speed, no net red-shift will be detected, even if both travel at a tremendous speed relative to some other object.

Tired Light Hypothesis

Modern astronomy use red-shift to calculate the speed with which distant bodies move, and since all distant objects display a red-shift, the conclusion is that our universe is expanding.

However, there are other ways to interpret red-shift in which a different conclusion can be reached.

The tired light hypothesis is one such interpretation. It attributes red-shift to a loss of energy by photons travelling across vast distances. The idea is that photons brush along dust and atoms on their way across the universe. They yield a tiny bit of energy to the environment they travel through.

The photons loose energy. As a result, they become redder.

The objection to this idea is that dust and atoms are so large that they will cause scatter. The fact that images from distant objects are sharp and crisp exclude the possibility of energy loss to such large particles.

However, if our universe is full of zero-point particles, then scatter is no longer a problem. Zero-point photons are too small to cause the much larger visible photons to scatter. But they will be receptive to a tiny bit of energy transfer.

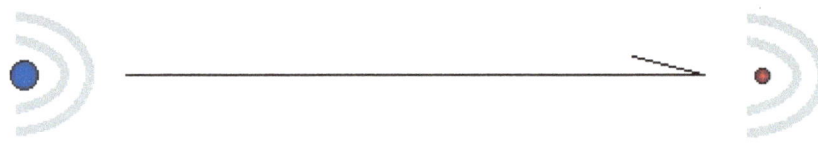

Photon loosing energy as it moves through the aether

The red-shift observed in astronomy may not after all be due to an expanding universe. It may instead be due to an aether of zero-point particles.

The background radiation attributed to the big bang may in turn be nothing but zero-point photons heated enough by visible light to give the universe a uniform glow.

If so, we no longer need to invoke a big bang or an expanding universe to explain the observed red-shift and background radiation. It can all be explained by the zero-point aether.

Intrinsic Red-Shift

Intrinsic red-shift is an idea proposed by the astronomer Halton Arp in order to explain the red-shift associated with quasars.

Quasars are young galaxies that appear to be physically linked to older galaxies. In some cases, a visible connection can be seen, like an umbilical cord, connecting the quasar to its parent galaxy.

However, the tremendous red-shift associated with quasars pose a great problem for those trying to interpret the visual data.

Quasar with red-shifted hydrogen spectrum, connected to parent galaxy

Using red-shift as a measure of distance, we end up with quasars being extremely distant.

This conflicts with their visible connection to galaxies that are estimated to be much closer to us.

Another problem with quasars is that they are relatively bright. If they are as distant as they appear to be based on red-shift estimates, they must be mindbogglingly energetic.

To get around this problem, Halton Arp proposed that quasars are made up of matter that is less massive than matter found in our solar system.

Hydrogen in quasars are lighter than hydrogen on Earth. The light spectrum of hydrogen in quasars is therefore red-shifted relative to the light spectrum of hydrogen found on our planet.

The red-shifts seen in quasars are not indicative of distance, but of age.

Matter starts out with little mass. As time progresses, mass condenses onto matter, making it gradually more heavy. Light from young matter is red compared to its older, heavier counterpart.

To see that this is a valid proposition, we can compare the light spectrum of hydrogen to that of its heavier isotope, deuterium.

Deuterium Blue-Shift

The chemical composition of distant objects can be determined by analysing their light spectra.

Atoms do not emit light in a continuous spectrum from red to blue. There are specific energies for which atoms absorb and emit light. The pattern is distinct and unique to each atom in the periodic table. It is the shape of this spectrum that reveals the chemical composition of a distant object.

If the light spectrum is shifted towards the red, we have what we call red-shift. If it is shifted towards the blue end, we have blue-shift.

An interesting thing to note is that the shape is determined by chemical properties only. The various isotopes of an atom all have the same shape. What distinguishes one isotope from another is its blue-shift. Heavier versions of an atom emit bluer light.

Hydrogen emitting redder light than deuterium

In the case of hydrogen, deuterium has a neutron attached to its nucleus. It is therefore twice as massive as regular hydrogen, which has a sole proton in its nucleus. Chemically, the two isotopes are identical. However, their light spectra can be used to tell them apart.

Deuterium can be identified by its blue-shifted spectrum. It is shifted towards the blue end of the spectrum by about 2 angstrom. This is how we know that the water in the tails of comets are rich in deuterium, and therefore not the source of water on our planet, which contains little deuterium.

All of this is easy to confirm in a laboratory. The heavier the atomic nucleus is for a given element, the bluer is its light spectrum.

Halton Arp's premise about light spectra and the mass of atoms is therefore correct.

Mass Condensation

Having established that the basic premise in Halton Arp's hypothesis is correct, we can take a look at what might cause matter to become more massive over time.

Halton Arp suggested that radiation in the form of high energy photons condense onto matter, thereby increasing the mass of matter over time.

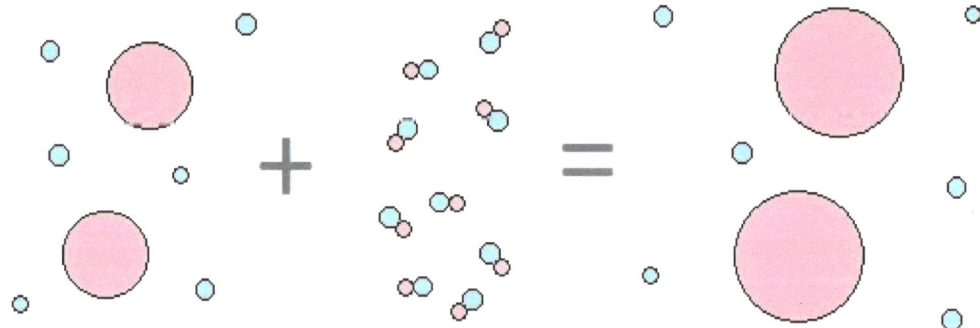

Small atoms + high energy photons = bigger atoms

Sticking with Morton Spears' particle quanta as our only building blocks, we see that the increased mass must be due entirely to a growing proton.

The electron consist of only 3 charged quanta, which corresponds to the negative orb of a photon.

The proton, on the other hand, consists of 2177 charged quanta. It consists of hundreds of photons.

It seems quite plausible, then, that the proton may have started out much smaller, and that it has grown to its current size over time.

The mechanism for this would be one in which protons sometimes consume gamma-ray photons as follows:

1. A gamma-ray photon crashes into a proton.
2. The photon breaks into an electron and a positron.
3. The proton consumes both the electron and the positron.

If this is how things work, then we can assume that the number 2177 for the current size of protons is an average. Some protons may be slightly larger, and others may be slightly smaller.

As time goes on, the size of the proton will grow. Gravity will become stronger as a consequence, and so will inertia.

Going back in time by studying the fossil records, we should find evidence of an environment in which gravity and inertia was less. As it happens, that's exactly what we find.

Darwin's Law

In his work on the origin of species, Charles Darwin postulated that no species will ever develop a size or shape that is grossly out of tune with its environment. Rather, the opposite will tend to happen.

Animals of completely different species will develop similar shapes and sizes given similar environments.

This law of nature is easy to confirm. There's no lack of examples:

The shrew looks like a mouse. The armadillo and the pangolin are similar to each other. The muskrat, beaver and coypu are also very similar.

Yet, none of these animals are very closely related. What they have in common is their environment and feeding habits.

These animals are similar in size and shape due to their similar environments.

What then are we to make of fossils in which we find a recognizable shape that is many times bigger than the same shape today?

Meganeura, lifesize model
(from Land of the dead blog)

The Meganeura was a dragonfly with a wingspan of 65 cm. It looked exactly like dragonflies look today, so its feeding habits must have been the same.

If inertia was as strong back in the time of the Meganuera as it is today, it would have constantly crashed into things. The sort of quick manoeuvring we see in dragonflies today would have been impossible for a dragonfly the size of the Meganeura.

Moreover, today's gravity would have made it impossible for the Meganeura to get off the ground. Its wafer thin wings would have broken if it tried.

The only way to explain the Meganeura is that it must have lived in an environment where both inertia and gravity were less.

Impossible Dinosaurs

Darwin's law is routinely ignored by those trying to make sense of the fossil records with the perspective that inertia and gravity were the same back in prehistoric times as they are today.

For example, the shape of the Tyranosaurus Rex tells us that it was a speed monster. Its enormous hind legs allowed it to run and jump. Most likely, it could turn on a dime by tossing its tail to the side if its pray tried to get away.

Tyranosaurus, by Marcin Polak from Warszawa / Warsaw, Polska / Poland
Tyranozaur RexUploaded by FunkMonk, CC BY 2.0,
https://commons.wikimedia.org/w/index.php?curid=31365817

Yet the official story is that it could hardly move.

This completely ignores Darwin's law. The shape of an animal never deviates grossly from its function. An animal with the shape of a speed monster cannot have lived a sloth-like existence.

The Brontosaurus, with its long swan-like neck is another example.

The official story on the Brontosaurus is that it carried its head straight out in front of it, and its tail was carried equally straight out behind it. It was a giant walking stick.

Brontosaurus: by MCDinosaurhunter
Own work, CC BY-SA 3.0
https://commons.wikimedia.org/w/index.php?curid=33465760

Using Darwin's law we would have expected the Brontosaurus to use its long neck to reach high up into trees. Its tail would be flexible and strong, allowing it to lean back on it as it stretched towards the highest branches.

However, with today's inertia and gravity, the Brontosaurus would have been unable to control its body in such a graceful way. The torque on its neck would have been enormous, and all blood would have drained from its head.

Even with its neck straight like a spear, problems remain. The fact that the Tyrannosaurus Rex could hardly move tells us that the Brontosaurus would have been incapable of any motion at all. The Brontosaurus was much bigger than the Tyrannosaurus. Its legs would have been crush by its own weight.

The Brontosaurus would have been an impossibility with today's inertia and gravity.

The Quetzalcoatlus

Even more impossible than the Brontosaurus is the Quetzalcoatlus. It was a dinosaur the size of a giraffe. Its neck was equally long as that of a giraffe, and it walked about on four legs.

Quetzalcoatlus
(c) M. Witton via G. Trivedi

Yet it could fly! When spreading out its front legs and fifth finger, it unfolded wings with a total wingspan of 16 meters.

The fact that an animal like that could fly is in itself baffling. However, that is not its only mystery. Its head, including its beak, was half the length of its giraffe-like neck. Instead of a tiny head perched on top of its long neck, like that of a giraffe, it had an enormous head with a giant beak.

Making this all the more mysterious is the fact that these flying monsters were predators. They could catch prey with the tip of their beaks. That would produce a tremendous torque on their necks. It seems that an abrupt movement of their heads would in itself have been sufficient to snap their necks.

Yet, they could hunt and catch prey with ease. They must have been able to move their heads about swiftly and forcefully.

The only way this could be possible is if the inertia associated with their heads was less. With less inertia, there would be less torque. Only a world where everything was substantially less massive would allow an animal like the Quetzalcoatlus to exist.

The Quetzalcoatlus is pretty much definite proof that both inertia and gravity have increased over time, exactly as hypothesised by Halton Arp.

Atomic Nuclei

If the neutron is nothing more than a proton with an electron stuck to it, we have no need for a strong force to keep the atomic nucleus from falling apart. The natural affinity between protons and electrons is all that's required to keep things together.

This is not a new insight. The original models of the atomic nucleus did not invoke the neutron. Instead, it had atomic nuclei made up of protons with electrons as glue.

This model is called the proton-electron atom.

Using this model, the assembly of large atomic nuclei from smaller ones is a straight forward matter of clicking protons into stable configurations with electrons tucked in between them.

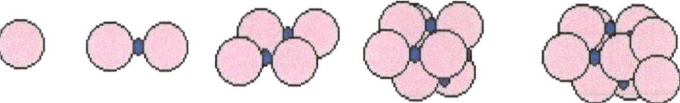

Atomic nuclei of hydrogen, deuterium, helium, lithium and beryllium

All of this fits well with the particle model laid out in this book. There is therefore no need to invent anything different.

Radioactivity

The proton-electron model for the atom has a fairly straight forward explanation for radioactivity in materials.

Large atomic nuclei are not big round balls with protons and electrons tossed together. They are more like crystals, with arms branching out and bending off to the sides.

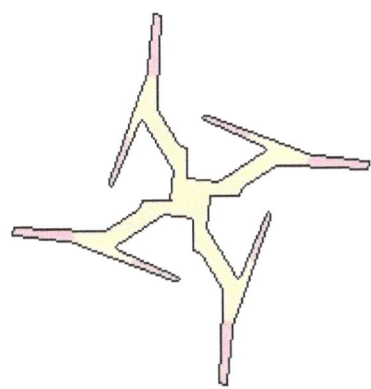

Large atomic nucleus

The core of these structures carry an overall neutral charge while the tips carry the positive charge.

This means that the tips of these structures repel each other.

As we go higher in atomic number, these structures start bending in on themselves. The branches get too close to each other, and the tendency is for these to break apart.

Atomic nuclei falling apart in this manner is what we refer to as radioactivity.

This problem does not arise suddenly at a certain atomic number. There are two radioactive materials with atomic numbers less than lead. These are element in which no stable configuration can be found despite their relatively small size. There is an arm that comes out in the wrong direction. For that arm to point away from its fellow arms, it needs to have one more or one less proton.

Beyond lead, there are no stable elements. They are all radioactive to some degree. The arms are turning in on themselves. There are no way to arrange them into stable structures.

Part of the problem is the relative size of protons and electrons. There is a whole lot of proton for each electron. There is a lot of mass to keep in place, and if mass condensation is real, as it appears to be, then this problem is likely to increase over time.

As protons grow in size, they become harder to keep in place inside the atomic nucleus.

Materials that are inert today may become radioactive in the future. Materials that are radioactive today may have been inert in the past.

The overall tendency is for matter to become more radioactive over time as mass condenses onto the proton.

Transmutations

A transmutation is a process in which an atomic nucleus goes from one position in the periodic table to another, usually one up or one down.

For example, potassium occupy the 19th position in the periodic table while calcium occupies the 20th position. The two elements are separated by a single charge. If a potassium nucleus drops a negative charge, as would be the case if it sheds an electron, it would no longer have a net charge of 19 but 20. It would have moved one up in the periodic table to become calcium.

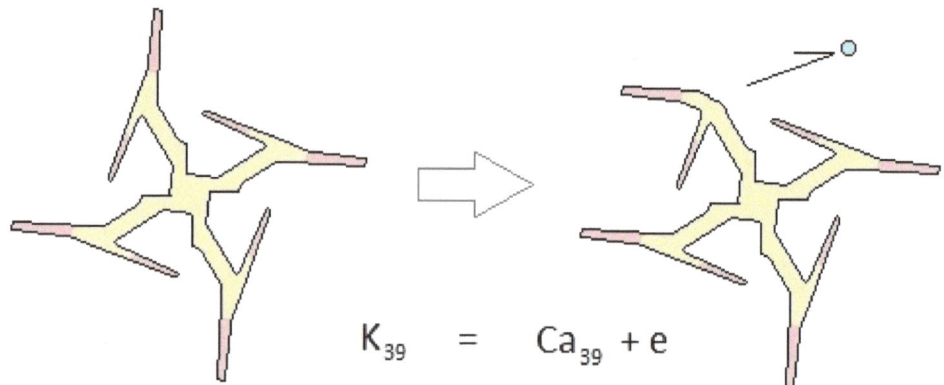

Potassium transmuting into light isotope of calcium by shedding an electron

Conversely, if a calcium nucleus consumes an electron, its net charge will go down from 20 to 19. This would move it one position down in the periodic table. It would be transmuted to potassium.

The mass of the atom before and after such a transmutation would be almost the same. They would only differ in mass by a single electron. This means that potassium would become a light isotope of calcium. Regular calcium would likewise become a heavy isotope of potassium.

All of this can be understood in terms of the proton-electron atom.

With the assumption that light is dielectric matter, a very similar mechanism can be added to this story.

Since a photon has the potential of becoming an electron-positron pair, we can in theory transmute matter up and down the periodic table by the use of light.

An atomic nucleus that consumes a positron without also consuming an electron will go one up in the periodic table in the same way that an atomic nucleus that consumes an electron without also consuming a positron goes down in the periodic table.

Potassium can in this way be transmuted into calcium. A high energy photon explodes into an electron-positron pair. The potassium nucleus consumes the positron, while letting the electron escape.

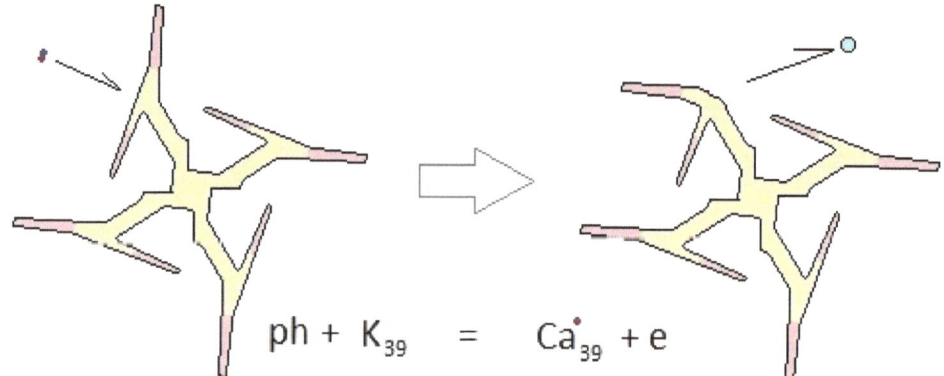

Potassium transmuting into light isotope of calcium by positron consumption

The calcium produced in this way will be heavier than the original potassium by the weight of a single positron.

This kind of transmutation can therefore be classified as a type of mass condensation.

Finally, we have the possibility of atomic nuclei transmuting through the consumption or shedding of a proton. This kind of transmutation is what we normally refer to as nuclear fusion and fission.

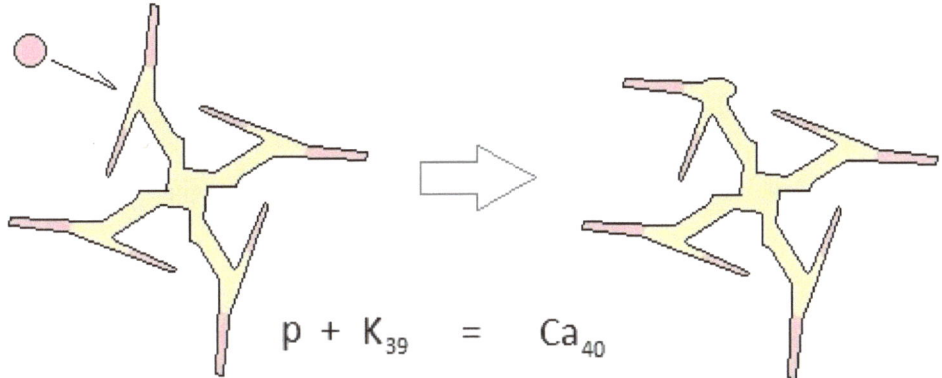

Potassium transmuting into a regular isotope of calcium through fusion with a proton

Potassium can be transmuted into calcium through fusion with a proton, and calcium can be transmuted into potassium through fission.

Antimatter

When an electron and a positron collide, we get what's called an electron-positron annihilation. The electron and the positron disappear, and a gamma-ray photon takes their place.

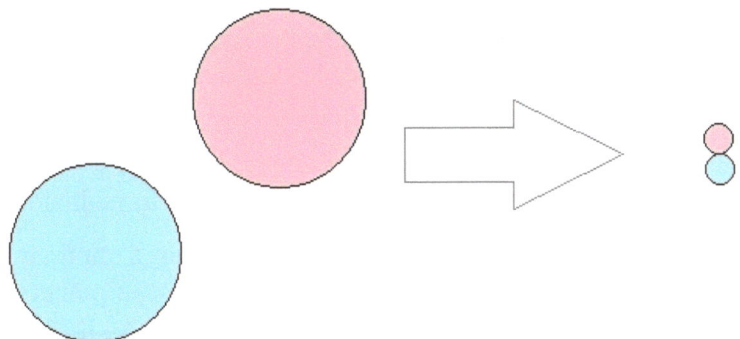

Electron-positron annihilation yields one gamma-ray photon

This was briefly touched upon in the chapter about electron-positron pair production.

In the physics laid out in this book, matter is never created nor destroyed. It is merely transformed. In the case of electron-positron pair production and annihilation, matter is being converted between the open and closed state.

Positrons are highly reactive. They combine readily with electrons to form gamma-ray photons. However, this does not prevent them from reacting with other particles as well. In the absence of an electron, the positron can react with atomic nuclei. It can produce transmutations and mass condensation.

Conventional theory, on the other hand, sees something entirely different in electron-positron pair production and annihilation. It sees the creation and destruction of matter.

This line of thinking has been extended to all kinds of matter, and has given rise to the idea of antimatter.

For every bit of conventional matter, there exists an antimatter. There's the anti-proton and the anti-neutrino. There are even anti-atoms made up of anti-particles in the same way that regular atoms are made up of regular particles.

Excited about this idea, people have tried to produce these anti-particles in laboratories, and they have succeeded, at least to a certain extent.

Various types of antimatter have been produced, and true to theory, these bits of antimatter always disintegrate into gamma-ray radiation when in contact with regular matter.

However, all of this can be explained just as well with the theory laid out in this book.

What has been produced are merely extremely unstable particles that naturally disintegrate into gamma-rays when in contact with regular particles.

As for detectable neutrinos and anti-neutrinos, these are highly charged neutrinos with sufficient energy to be detected. When a positively charged neutrino hits a negatively charged neutrino, they cancel out their charges, and their energy is transferred to a nearby zero-point photon.

The fact that highly unstable particles can be created in a lab, and neutrinos can be excited to such an extent that they are detectable, does not prove the existence of antimatter. It merely demonstrates that it is always possible to produce such particles, given enough energy to do so.

Hollow Planets

We know from measuring the electric potential gradient of our atmosphere that our planet is negatively charged relative to the ionosphere. The potential difference is about 300,000 volt.

It is the potential difference between the ionosphere and the surface of our planet that keeps our atmosphere from escaping into space. The much weaker gravitational force would not be able to do this on its own.

The negative charge on the surface of our planet is most likely matched with a corresponding positive charge at its centre. This would mean that there is a repelling electrical force inside Earth.

Since gravity is measured from the centre of astronomic bodies, and not from their surfaces, as is the case with the electrostatic force, there can be no net gravity at the centre of planets, moons and stars.

This means that there is nothing to prevent astronomic bodies from being hollow. There is no force at the centre of such bodies to counter the effect of internal electric repulsion. Nor is there anything to counter centrifugal forces due to spin.

If a cavity was to develop inside an astronomic body, there would be no way to make it disappear.

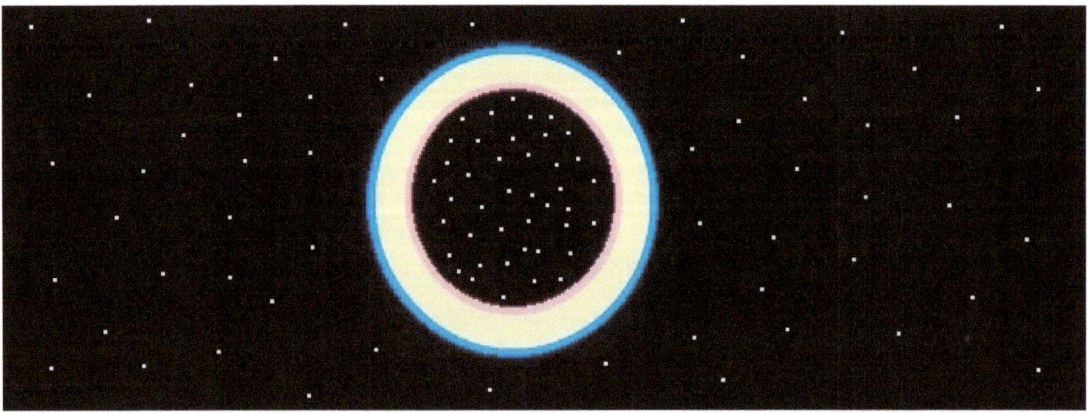

Cross section of a hollow planet with repelling electric force at its core

This was first recognized by Isaac Newton in his mathematical work on gravity. In his shell theorem he demonstrated that there is nothing to stop astronomic bodies from developing empty cavities.

When the astronomer Edmond Halley suggested to Newton that our planet may be hollow, Newton did not object. There was nothing in Newton's theory to counter Edmond Halley's suggestion.

Now that we know that there most likely is a strong repelling force inside all astronomic bodies, there is even less reason to object to such a notion.

The enormous pressure that undoubtedly exists at the core of all astronomic bodies is no argument for a solid core. The pressure inside the walls of a tunnel does not make tunnels collapse. The same is true for any cavity inside our planet.

Seismic evidence for a solid core is often used as an argument. However, reconciling seismic data with a solid core is extremely difficult. Using a hollow Earth model is much easier. Jan Lamprecht demonstrated this in his work on the subject.

Even NASA appears to accept the possibility of hollow planets. According to their own measurements, Jupiter's core is considerably less dense than its outer layers.

The only serious objection to a hollow Earth model is the fact that gravity at the surface of our planet indicates that it must be made up of something extremely dense.

The latest estimate is of a super-dense crystal at Earth's core. This material, which only exists in theory, and no-one has ever been able to produce in a laboratory, has all sorts of fantastic properties. This is all required in order to reconcile observed seismic and gravitational data with current theory.

However, there is a simple way around this. By recognizing that our planet is a gigantic charged capacitor, we can make the proposition that the dielectric material inside capacitors will add to the gravitational force when sufficiently charged.

Gravity and Capacitance

If the gravitational force is communicated by neutrinos in the manner suggested in this book, then this force is dependent on four factors:

1. The distance between two objects
2. The total number of charged quanta involved (mass)
3. The availability of neutrinos (gravitational constant)
4. The magnitude of the in-print communicated by each charged quantum

The first three factors are all included in Newton's universal law of gravity. What is not included is the possibility that charged matter may be different than neutral matter in respect to the gravitational force.

This relates to point 4. The magnitude of the in-print communicated by neutrinos may be dependent on the electrical environment in which the in-print is made.

Since our planet has a charged surface, it is by definition a capacitor. Most likely, our planet is fully charged, which would make the total charge carried by our planet truly enormous.

This charge exerts stress on the crust of our planet. Positive quanta are pulled towards the negative surface. Negative quanta are pulled towards the positive surface.

Since both protons and electrons are dielectric, an internal stress develops. The negative hoops and the positive hooks of these particles get pulled further out than normal.

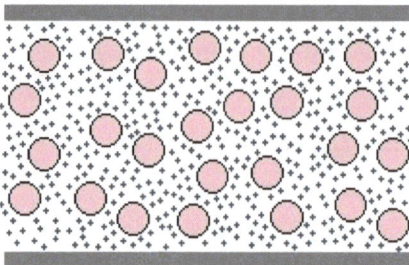 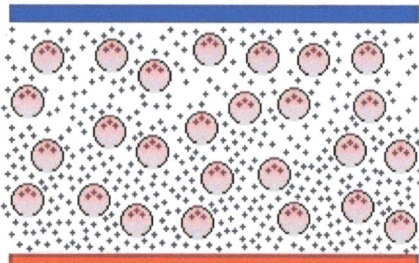

Uncharged and charged capacitor

When neutrinos hit these stressed particles, they get in-printed with a more pronounced footprint. This in turn translates into more forceful collisions between neutrinos in the surrounding space, and therefore stronger gravity.

Expanding Planets

Assuming that an increase in radioactivity is a side-effect of mass condensation, and that mass condensation is real and ongoing, all astronomic bodies will experience an increase in internal pressures over time.

This is because radioactivity results in an increase in the number of atoms in a given space. Where there was once only one atom, there are suddenly two. With more atoms occupying the same space, pressure builds up.

A planet may stay unchanged in size for a very long time. However, at some point, the internal pressures will become so great that it will crack and start to expand.

From evidence available to us, it appears that our own planet stayed pretty much unchanged in size up until about 300 million years ago. Our planet had at that time a diameter roughly half of what it has today.

However, ever since then, our planet has been expanding.

This is based on the fact that continental crusts are about 4000 million years old, while no ocean floor is older than 300 million years.

Also, if we cut away all the oceans on our planet, the continents fit perfectly together onto a sphere half the diameter of present day Earth.

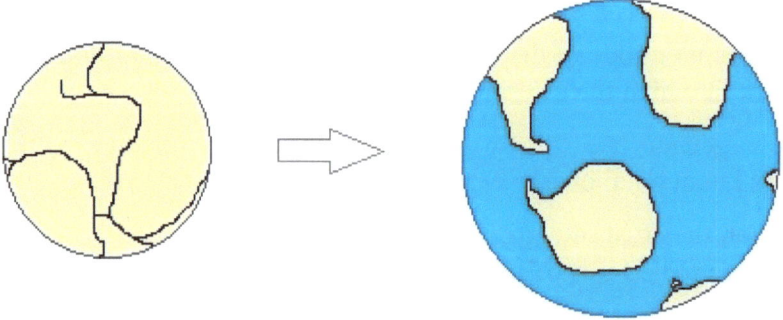

South pole view of the expanding Earth

Oceans are rifts produced by the expansion of our planet, while continents are the original crust.

It appears then, that mass condensation can explain both the size of the dinosaurs and the expansion of our planet.

Gravity Anomalies

The gravitational force is not equally distributed across our planet. Some places have more gravity than others. This is true, even when measurements are adjusted for height above sea level and the centripetal force of our spinning planet.

These gravity anomalies are not randomly distributed. They coincide with geological activity. Places with a lot of geological activity have stronger gravity than areas that have little geological activity.

It does not matter if the geological activity is due to uplifting of mountains, or formation of rifts.

Iceland, situated on the mid-Atlantic rift, has stronger gravity than normal. The same is true for the Himalayas and Andes where there has been a lot of uplifting.

North-east Canada and Tibet have very little geological activity, and they both have relatively less gravity than other areas.

Conventional theory holds that mass alone is the source of the gravitational force. The anomalies are therefore explained by a greater abundance of especially dense matter in the geological active zones. Dense matter floats up through less dense matter in both regions of rifting and uplifting.

However, this theory violates the law of buoyancy. Dense matter sinks. It never floats upwards. Uplifting should therefore result in less gravity, and the same should be true for rifting. In both cases, light matter should bubble up towards the surface.

But if gravity is due to capacitance as well as matter, the mystery of gravity anomalies solves itself. Especially if our planet is hollow.

All else being equal, the capacitance of a thin capacitor is greater than the capacitance of a thick capacitor. An expanding hollow planet would therefore be increasing its capacitance, and this would be especially noticeable in areas where the capacitor is cracking.

If the role of capacitance as a source of gravity in our planet is greater than the role of inertial mass, then surface gravity will increase with expansion. The reduction in overall density due to a thinner crust will be made up for by greater capacitance.

An expanding planet will display two types of cracks. There will be rifts where the old crust is pulled apart, and there will be mountains where the old crust breaks in order to fit onto the larger sphere. In both cases, we end up with a thinner crust along the cracks than in areas where there is no cracking. The geologically active areas will have more capacitance, and therefore more gravity than the geologically inactive areas.

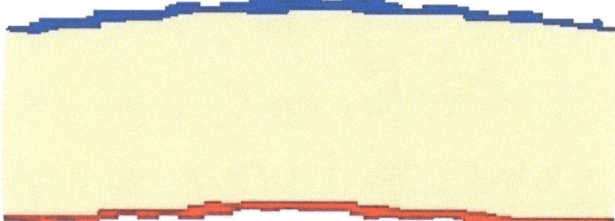

Geologically inactive Tibet and north-east Canada have thick crusts

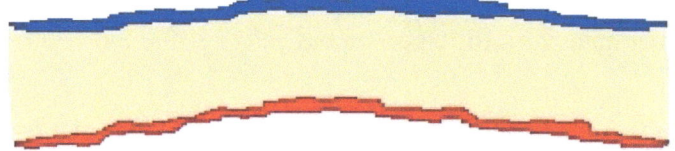

Rift zones like Iceland have thin crusts

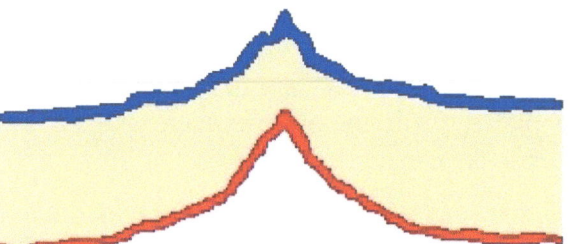

Uplifting cracks like the Himalayas and Andes have thin crusts

It appears then that our planet is a charged body that is both hollow and expanding.

Origin of Water in the Oceans

If all ocean floors are newly formed due to expansion, then we need to explain how these enormous rifts have been filled so perfectly with salt water. The amount of water is neither more nor less than what has been required to fill the rifts.

Ocean-centric view of Earth

By Serg!o - File:Oceans.png, Public Domain, https://commons.wikimedia.org/w/index.php?curid=11691840

The only way this could be the case, short of a miraculous coincidence, is that the water has come from inside our planet. If it came from outside our planet, the coordination between expansion and water supply would be impossible. There is no way the heavens could be coordinated in such a way that they supply water at the exact rate of planetary expansion.

Furthermore, comet tails have water rich in deuterium. They are therefore not the source of water on Earth.

The abundance of salt in our oceans is further evidence that the water cames from within our planet which has huge salt domes hidden deep below its crust.

If the expansion of our planet is due to radiation, as suggested in this book, then water may even be synthesized inside our planet as a part of the expansion process. Heavy elements split off hydrogen and oxygen atoms as they decay through radiation. The result is water.

From observing rifts and volcanoes, water vapour appears to be an abundant component of their venting. All evidence point to Earth as the source of salt water in the oceans.

Stability of Orbits

Electric repulsion due to similar charged surfaces is what keeps orbits from collapsing or flying apart at the slightest disturbance.

To see how this works, consider our Moon and what would happen if some external force were to push it hard towards our planet.

Without electric repulsion, our Moon would speed up, the force of gravity would be stronger on average, and the orbit would be elongated. Pushed hard enough, our Moon would crash into our planet.

Conversely, if the push was away from us, our Moon would start following a wider orbit, also more elongated than it is today.

However, as soon as we include the effect of electric repulsion, we see that things will quickly stabilise.

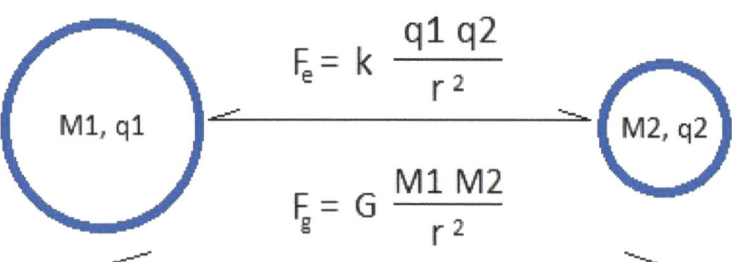

Electric repulsion and gravitational attraction

Contrary to the gravitational force which is calculated from the centre of objects, the electrical force is calculated between the surfaces of objects.

This means that electrical repulsion increases more quickly than the attraction of gravity for bodies that approach each other. It also means that electric repulsion decreases more quickly than gravity for bodies that move apart.

The net result of this is that we get a buffering effect. If our moon is pushed towards us, repulsion kicks in. If our moon is drawn away from us, repulsion decreases more than attraction. Oblong orbits are thereby restored to near perfect circles.

This is true for all astronomical bodies and the reason why collisions between such bodies are rare.

Furthermore, we can make the prediction that if an expanding planet is changing its gravity primarily due to an increase in charge, orbiting moons will be pushed farther away. The increase in charge will trump the corresponding increase in gravitational mass.

As it happens, our Moon is receding from us by a few centimetres a year. This is attributed to tidal forces. However, it may also be due to ongoing changes in our planet's total charge.

All orbits may be changing over time, with young orbits being generally closer together than older ones.

Meteorites

When a meteorite enters Earth's atmosphere, it soon starts to glow. This goes on for a short while before it vanishes, either quietly, or in a flash.

Conventional theory claims that this is due to friction between the meteorite and the atmosphere. However, the magnitude of the explosions observed when large meteorites enter the atmosphere makes it hard to believe that this is all due to heat convection. The difference in electric potential between our atmosphere and the incoming object is a more likely source of such enormous energy.

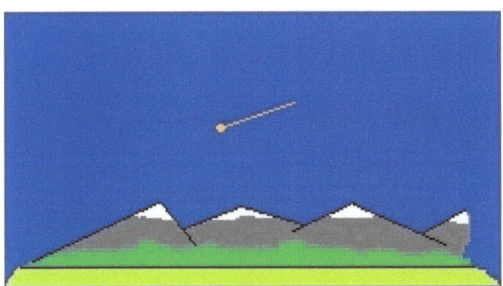 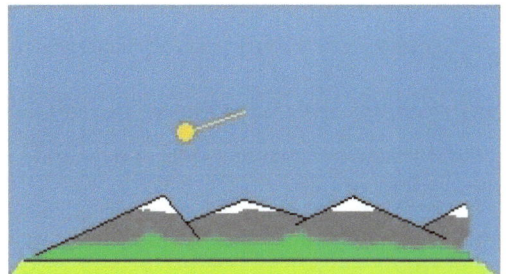

Large meteorite glowing before exploding

The glow and the explosions associated with meteorites are most likely electrical discharges. Small objects manage to equalize their electric potential with the atmosphere by sparkling brightly. Larger objects explode.

Impact Craters

Most meteorites, when we see them in the night sky, come in at an angle. Rarely do we see them come straight down from above.

This mean that such bodies would leave oblong craters if they were to strike the crust of our planet.

However, all impact craters are circular. There is not a single oblong impact crater on our planet, nor is there any such crater on the Moon.

This suggests strongly that all meteorites explode before they hit ground. Only tiny objects can make it all the way to ground without exploding.

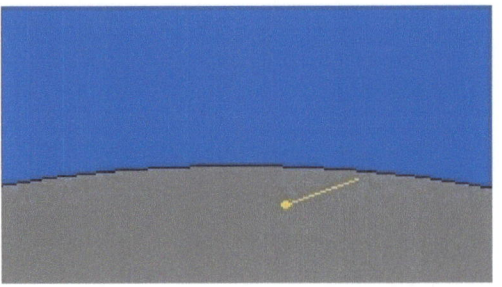 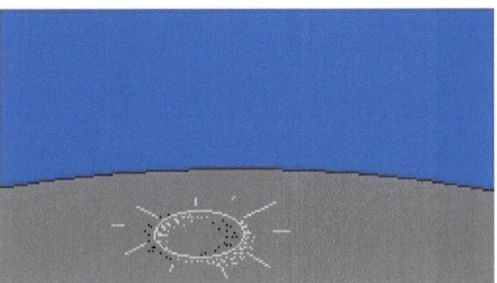

Impact crater produced by exploding meteorite

This is true on Earth, on the Moon, and everywhere else in our solar system. Since there is no atmosphere on the Moon, we must assume that these explosions are electrical in nature.

Moon Craters

If craters on the Moon are solely due to impact, as many believe, then we should expect craters to be randomly distributed. Some shielding from Earth would be expected, but for the rest, the craters should appear with no clear pattern.

However, this is not the case. Small craters are predominantly located on peeks and ridges. It is quite common to see them on the edge of older, bigger craters, or lined up neatly along a ridge.

 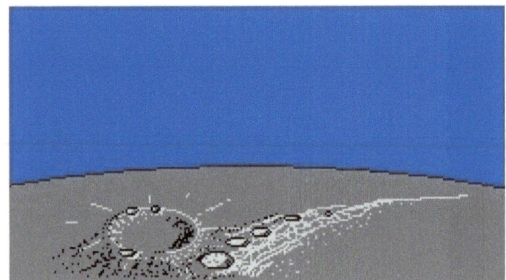

Electrical cratering on exposed edges

Larger craters are uniformly spread out. Viewed from the north pole or south pole, a spiralling pattern of large craters can be seen.

All of this indicate some sort of continuous process in which craters are excavated slowly over time.

The prime suspect in this case would be ion winds. Charged particles move along the surface of our moon until they find a suitable escape point, usually on a ridge or other high point. They spiral around the escape point a few times before leaving the surface.

Over time, craters appear, evenly spaced out, themselves forming a spiralling pattern.

Lunar north pole

By NASA/GSFC/Arizona State University - http://wms.lroc.asu.edu/lroc_browse/view/npole (see also http://photojournal.jpl.nasa.gov/catalog/PIA14024), Public Domain, https://commons.wikimedia.org/w/index.php?curid=31697472

The craters on the Moon are not proof of a violent past, but mostly the result of dust and other particles fluttering along its surface.

Impact craters are relatively rare in comparison to electrically excavated craters.

Comets

If the voltage potential of our atmosphere is a few hundred thousand volts, then something similar should be true for the solar system as a whole. The electric potential between the inner and outer solar system should be enormous. An object moving from the outer to the inner regions, or visa versa, should experience electrical stress similar to that experienced by meteorites entering our atmosphere.

Comets, with their oblong orbits around our Sun, should display evidence of electrical activity, which is exactly what they do.

Long before comets enter regions warm enough to melt water, they develop long tails rich in water.

However, comets are not icy bodies. They are rocks. Space probes that have observed comets up close, and even landed on them, have found no source of water, only barren rock and dust.

Comet 67P in January 2015 as seen by Rosetta's NAVCAM
By ESA/Rosetta/NAVCAM
https://www.flickr.com/photos/europeanspaceagency/16456721122/, CC BY-SA 2.0,
https://commons.wikimedia.org/w/index.php?curid=40847079

The electrical explanation for this is that the water observed in the tail of comets is synthesized through nuclear fission. Atoms of heavy elements are ripped apart by electric stress. Oxygen and hydrogen is in this way synthesized.

This also makes the abundance of deuterium in comets' tails easier to understand. Heavy elements have a larger proportion of neutrons in their nuclei than lighter elements. Ripping heavy elements apart would result in a general abundance of heavy isotopes in the elements produced.

Rogue Planets

Just like comets, roaming the universe, there are planets that do not belong to any solar system.

When such planets enter an established solar system, forces are unleashed to either capture it or to eject it permanently from the system. This can easily be understood in terms of what has already been said about orbits, meteorites and comets.

First thing to note is that a rogue planet, although gravitationally attracted to other planets, and the central star, will have an electrically charged surface that makes direct collisions highly unlikely.

We don't have to worry about rogue planets crashing into Earth. However, that is not to say that a close encounter with such a planet would be entirely harmless.

At the very least, there will be a very strong ion wind associated with such an event. This will cause severe storms on Earth, way worse than anything anyone alive has ever experienced.

If the encounter is sufficiently close, there will be a discharge between the two planets as they seek to equal out their charge difference. This will be extremely destructive, wiping out all life wherever the discharge hits. A valley will be carved into each of the two planets as the discharge moves along their surfaces.

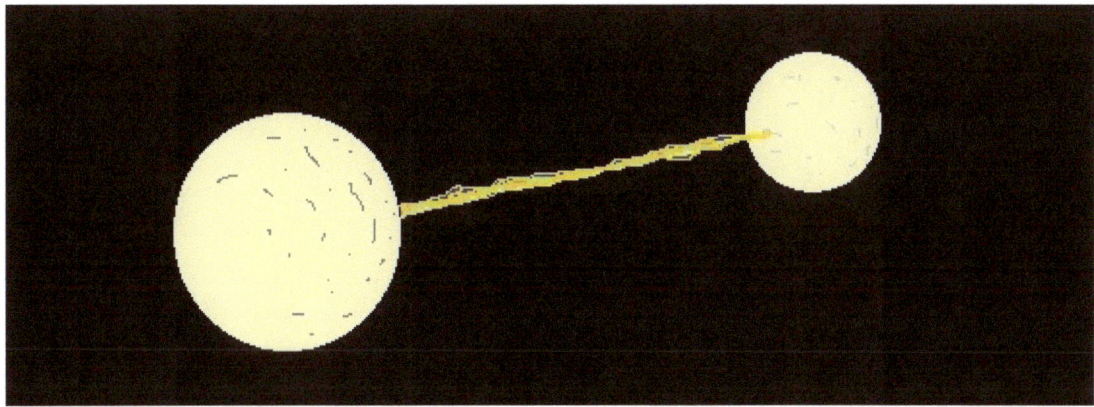

Electric discharge between two planets

The net result of this will be repulsion. The rogue planet will either be ejected from the solar system or captured by it.

If captured, the newcomer will push and jockey for position. It will seek an orbit in harmony with all the other planets. This may require other planets to change their orbits to allow for the newcomer.

There will be chaos, but it will be relatively short lived. The combination of electrical repulsion and gravitational attraction acts like a shock absorber. The rogue planet will be rained in. It may not take more than a few decades to integrate the newcomer into the solar system.

After a few thousand years, it will be as if the rogue planet has always been a member of the family. However, some evidence of a violent past will remain.

Valles Marineris

Mars has a scar. It's called Valles Marineris.

Mars with Valles Marineris clearly visible

By NASA / USGS (see PIA04304 catalog page) -
http://nssdc.gsfc.nasa.gov/photo_gallery/photogallery-mars.html
http://nssdc.gsfc.nasa.gov/image/planetary/mars/marsglobe1.jpg, Public Domain,
https://commons.wikimedia.org/w/index.php?curid=19400

This scar is either a rift due to planetary expansion or the result of a prolonged electrical discharge between itself and another planet. Which one of the two it is can be determined by taking note of certain tell tale features.

The edges of the scar have the characteristic zigzag pattern that electricity produces.

The scar is widest in the middle by quite a lot. This is also where a number of smaller scars are formed, indicating a widening out of the current flow at the moment the two planets were the closest together.

The scar is wide all the way. It does not taper into a very fine line, which we would expect from planetary expansion.

To the left in the picture, we can see a round pattern. The discharge has been lingering at this point as the two planets moved away from each other. This is what discharges do. Once an arc has been established, the connection is not immediately broken by pulling away. It sticks.

Valles Marineris can therefore be taken as evidence of a rogue planet that once roamed our solar system. This rogue planet may have been Mars itself. It may also have been some other planet, possibly one that was successfully ejected by Mars.

It appears then that some great battle took place in the heavens in some distant past. It may not be entirely coincidental that Mars is known as the god of war.

Grand Canyon

Here on Earth we have the Grand Canyon, our own mini-version of Valles Marineris. Seen from space, it looks like an electrical scar.

Satellite picture of Grand Canyon

By Erthygy - Own work, CC BY-SA 4.0,
https://commons.wikimedia.org/w/index.php?curid=66479110

The official explanation for it is that it has been carved out by water trickling through it over millions of years. However, it does not look like any other valley anywhere else on Earth. No other river has carved out an electric scar shaped valley.

If we stick with our hypothesis that there has been a rogue planet in our solar system, it seems more likely that the Grand Canyon is the product of a close encounter with this planet. The same planet that caused enormous damage to Mars came dangerously close to Earth as well.

If the Grand Canyon was created through discharge between Earth and a rogue planet, the entire canyon may have been carved out in less than an hour.

Such an enormous event is hard to comprehend. It is hard to even begin to imagine the power required to perform such an act of destruction in such a short time. However, we are not talking about a meteorite or a comet. We are talking about an object the size of a planet.

For perspective, we can look up industrial capacitors on YouTube to see what sort of damage such devices can do. A capacitor the size of a beer keg can easily vaporize bits of a coin or a pebble. It can blow watermelons to bits. All sorts of fun can be had.

A capacitor the size of a planet can without doubt do some serious damage to other planets. If the planet was sufficiently large and charged, it could even have blown up a planet or two on its destructive way through the solar system.

Exploding Planets

Capacitors are known to explode when charged too much. This means that if a highly charged planet comes in contact with a smaller planet, the bigger planet may cause the smaller one to explode.

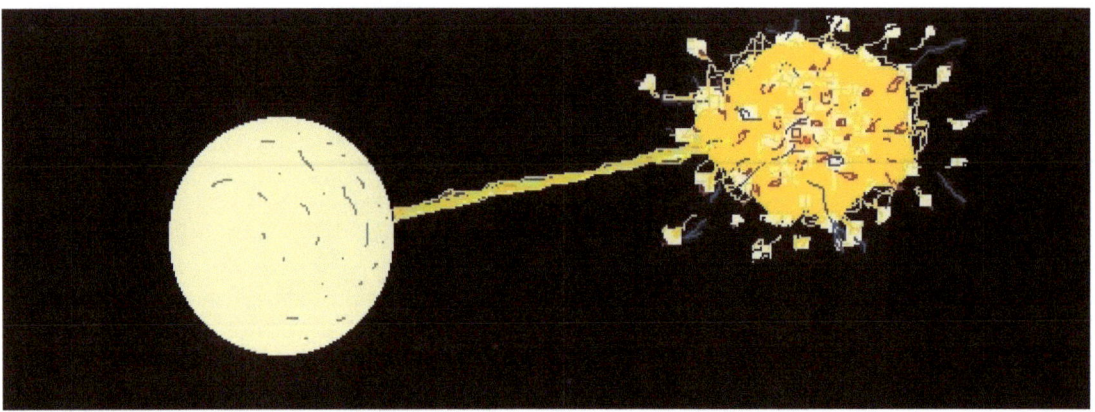

Rogue planet blowing up a smaller planet

The effect would be like connecting a fully charged industrial capacitor to a much smaller capacitor.

Again, we are struck by the enormity of such an event. It seems impossible. However, the laws of electricity scale perfectly. What is true for capacitors in laboratories on Earth is also true for planet size capacitors in space. Charged too much, they explode.

Not only is such an event a theoretical possibility. There is evidence to suggest that it has happened in our own solar system, quite possibly due to the same rogue planet that scarred Mars and zapped the Grand Canyon into the crust of our planet.

Ceres, Phaeton and the Asteroid Belt

Between Jupiter and Mars, lies the asteroid belt. It is a large collection of rocks of various sizes that orbit the Sun together with the dwarf planet Ceres.

Ceres

By Justin Cowart - Ceres - RC3 - Haulani Crater, CC BY 2.0, https://commons.wikimedia.org/w/index.php?curid=49700320

Ceres is not much of a planet. It is a good deal smaller than our own moon.

Current theory holds that the asteroid belt is left over rubble from the creation of our solar system. Too much gravity from Jupiter prevented the successful creation of a proper planet, so all we got was Ceres and a bunch of unused building material.

However, there is an older theory that tells quite another story. This theory harks back to the ancient Greeks, and was the accepted theory up until the 20th century. In this theory, a planet called Phaeton was destroyed in a squabble with Jupiter.

The discoveries of Ceres and the asteroid belt by 19th century astronomers were taken as proof that Phaeton had indeed existed. Ceres was either a large part of Phaeton or it was its moon.

This older theory fits very well with the hypothetical rogue planet. On its way from Jupiter into the inner solar system, it blew up Phaeton, scared Mars, and zapped Earth.

A trail of destruction leads us to Jupiter. The closer we get to the gas giant, the more monumental is the destruction observed.

Jupiter's Children

There is a lot of energy associated with Jupiter. Everything about it is colossal. It rotates faster than any other planet. There are strong winds and enormous storms. Its famous red spot is a storm the size of a planet that has raged for centuries. The whole planet is under intense stress, and the way it alleviates this is by spinning fast and generating storms.

However, these two actions may not always be sufficient. If Jupiter comes under sufficient stress it must find a third way to rid itself of surplus energy.

An effective way to do so would be to shed some of its atmosphere.

As we have seen, meteorites explode when under sufficient electrical stress. Comets shed matter by growing a tail. This is how electrical stress is alleviated. Matter is ejected from the stressed body.

The way Jupiter will do this is by first producing a big storm, rich in minerals. This storm will have a dark brown or red colour. It may last for centuries and it may never be ejected. However, under sufficient stress, Jupiter will eject the mineral rich storm.

The famous red spot on Jupiter is not just a storm. It is an embryonic moon.

This embryo can either return into nothing, absorbed by Jupiter itself, or it can compact into an intensely hot and charged ball the size of a moon or a planet.

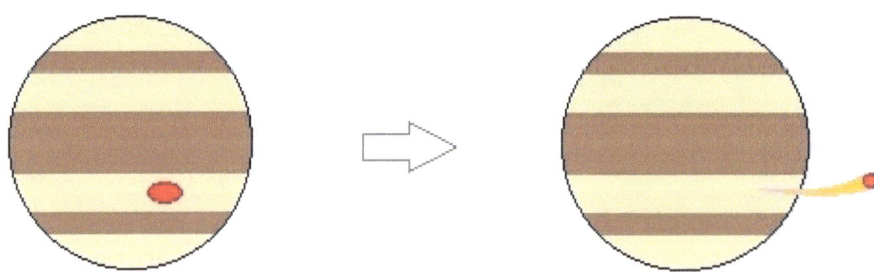

Large storm ejected from Jupiter in the form of a moon

At the moment, the red spot on Jupiter is becoming smaller in diameter, and at the same time taller. If this continues, Jupiter will give birth to yet another child.

Should that happen, we must hope that the birth is relatively uneventful and that the child quickly settles in among its siblings as yet another moon of Jupiter, because a white-hot and extremely charged body emanating from Jupiter is a very accurate description of the rogue planet we have been looking for.

Putting all the evidence together, we get the following description of the fateful events that led to the destruction of Phaeton.

Already under considerable stress, Jupiter was antagonized by the smaller planets closer to the sun. They lined up in the direction of Jupiter, allowing for a freer flow of energy from the Sun to Jupiter.

The additional energy provoked the birth of a planet with sufficient momentum to escape the gravitational pull of Jupiter. The new planet raced towards the Sun, following the electric current set up by the planetary alignment.

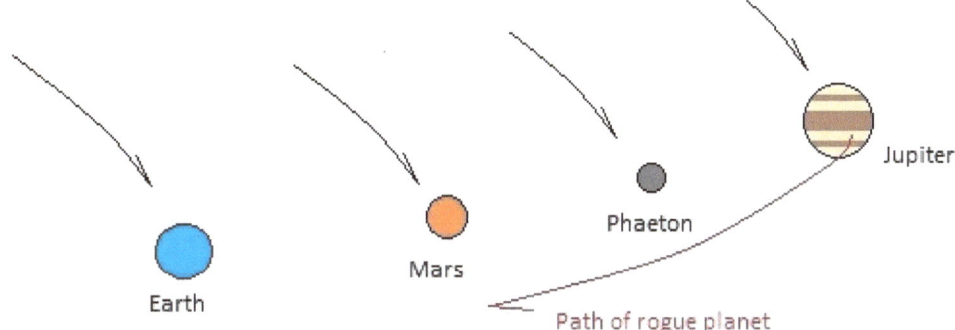

Planetary alignment and path of rogue planet

The rogue planet was on a collision course with Phaeton, Mars and Earth.

Unloading the bulk of its charge on Phaeton, it obliterated it. On its way passed Mars, it scarred it badly. By the time it reached Earth, it had unloaded most of its charge. Yet, it was still able to zap Grand Canyon into our planet's surface.

Jupiter's Daughter

The rogue planet that once roamed our solar system appears to have been a stray child of Jupiter.

According to Greek legend, that child is Venus. Born from a storm, and able to throw punches like her father, she caused all sorts of trouble. Beneath her brilliant exterior hides a wild child with a mind of her own.

Could it be that the Greeks had it right? From what we have been able to deduce about the rogue planet, it certainly looks that way. The planet we are looking for is a relatively large rock planet, most likely with a thick atmosphere. That sure sounds like Venus.

Taking a closer look at Venus, we notice something odd.

She's rotating very slowly, and she's rotating the wrong way.

All planets in the solar system rotate the same way. Their speed of rotation is related to their size and energy level. Big planets with thick atmospheres spin quicker than smaller, rocky planets.

But Venus is hardly spinning at all. In fact, she's slowing down. She's being slowed down by the Sun and ordered to spin the other way. It is as if Venus is a newcomer to the solar system, still trying to learn the rules of the game.

But most telling of all is what Venus is hiding under her thick atmosphere. There are all sorts of scars on her surface, as if she's been in several fights with other planets.

Global radar view of Venus (without the clouds) from Magellan between 1990 and 1994

By NASA - http://photojournal.jpl.nasa.gov/catalog/PIA00104, Public domain, https://commons.wikimedia.org/w/index.php?curid=11826

The scars are softened and rounded off by the heat and the acid atmosphere, but they are still visible.

It appears that we've found our culprit. Venus was the rogue planet that caused so much damage to our solar system, and she really is the daughter of Jupiter, as Greek mythology has it.

The Electric Universe

In the physics laid out in this book, gravity takes a side-role to electricity. Gravity is due to a tiny imbalance in the electrical force. It is of little importance in monumental events, such as those described in the chapters above.

Gravity is only a significant force when there is electric stability. It is therefore a mistake to assume that all we see in the universe is primarily due to gravity.

This is not a new idea. Kristian Birkeland recognized the importance of currents in space as early as the late 19th century. He explained both the auroras and Saturn's rings in terms of electricity. In his terella experiment, he reproduced Saturn's rings in his laboratory.

Kristian Birkeland and his terrella experiment

Public Domain, https://commons.wikimedia.org/w/index.php?curid=307997

Following in Kristian Birkeland's footsteps, Hannes Alfvén received the Nobel Prize in Physics in 1970 for his work on magnetohydrodynamics, one of his many theories related to electric plasma in space.

The idea that Venus is a relatively new body emanating from Jupiter was first suggested by Immanuel Velikovsky in his book Worlds in Collisions, published in 1950. Velikovsky's version is more elaborate and reliant on ancient myths than the version presented in this book. However, the basic premises and conclusions are the same.

Inspired by Velikovsky's work, Ralph Juergens proposed an electric model for the Sun in 1972.

Today, Wallace Thornhill and David Talbott are the main proponents of the idea that the universe is driven by electromagnetic forces. Through their Thunderbolt Project, they have produced a wealth of easily accessible material that they have made readily available on the web.

Many others have also come to the conclusion that electricity, rather than gravity is the driving force of the cosmos.

Currents in Space

In a picture taken by the Hubble Space Telescope, a jet of matter can be clearly seen ejected from the centre of a galaxy. The jet is 4,400 light-year long.

Plasma jet ejected by a galaxy

By NASA and The Hubble Heritage Team (STScI/AURA)
HubbleSite: gallery, release., Public Domain, https://commons.wikimedia.org/w/index.php?curid=102873

The fact that the jet hardly disperses over such a long distance suggests that it is highly charge. A strong magnetic field is required to keep something like this together over such a long distance, and the most likely source of that magnetic field is the jet itself.

Charged gases such as these are generally referred to as electric plasma. Their behaviours are different from electrical neutral gases. For one thing, they can keep together for enormous distances without dispersing.

Donald Scott, a contributor to the Thundrerbolt Project, has a very insightful lecture on this topic, worth looking up on the web for those interested in more information on this. In the same lecture, he discusses the mechanisms behind planetary formations.

Birth of Stars and Planets

There is no lack of evidence for plasma currents in space, and these currents come in all sizes.

There are the truly huge ones, stringing galaxies together like pearls on a string. Then there are the big ones that do the same for stars. Then there are the relatively small ones connecting planets to their central star. These are responsible for the auroras that we see in the atmosphere of planets.

The overall impression is that of a neural network with stars and galaxies forming the nodes and the plasma currents forming the synapses.

In all of this, there are the occasional bright flashes. These are the so called supernovas.

Standard cosmology attribute these flashes to the death of stars. However, Donald Scott suggests otherwise. He sees them as the births of stars and planets.

As we have already discussed regarding Jupiter and Venus, large planets with thick mineral rich atmospheres can give birth to smaller objects by ejecting a highly charged body the size of a planet or moon.

Stars can do this too. But when they do, the size of the object ejected is that much larger. Large stars can sweat off objects the size of gas giants or small stars.

This explains why binary star systems are relatively common, and why gas giants can be found very close to stars.

All of this is accompanied by bright flashes. However, only the brightest of them are categorised as supernovas, and to explain the most energetic flashes, something much bigger must be going on.

The biggest supernovas are most likely due to short circuiting of large plasma currents. The technical term for such a short circuit is a z-pinch, and it has the effect of pulling matter together.

A z-pinch can easily be produced in an electrical laboratory.

Pinched aluminium can, produced from a pulsed magnetic field

By Bert Hickman, CC BY-SA 3.0, https://commons.wikimedia.org/w/index.php?curid=28083081

As can be seen in the above picture, a z-pinch can crush an aluminium can. If the can had been made of something more fluid, it would have been crushed completely.

Keeping in mind that electricity scales very well from the very small to the positively enormous, we can now imagine a dusty, mineral rich plasma current with a diameter many times that of a solar system.

Such a current would be like an enormous cylinder with several layers of positively and negatively charged tubes nested inside each other.

In balance with itself, the current will be cold and invisible, only detectable by the fact that there is a star at each end of it.

However, should such a current short circuit, there would be an enormous flash, followed by a lingering glow.

The glow will be visible as an hourglass shape similar to the crushed aluminium can depicted above.

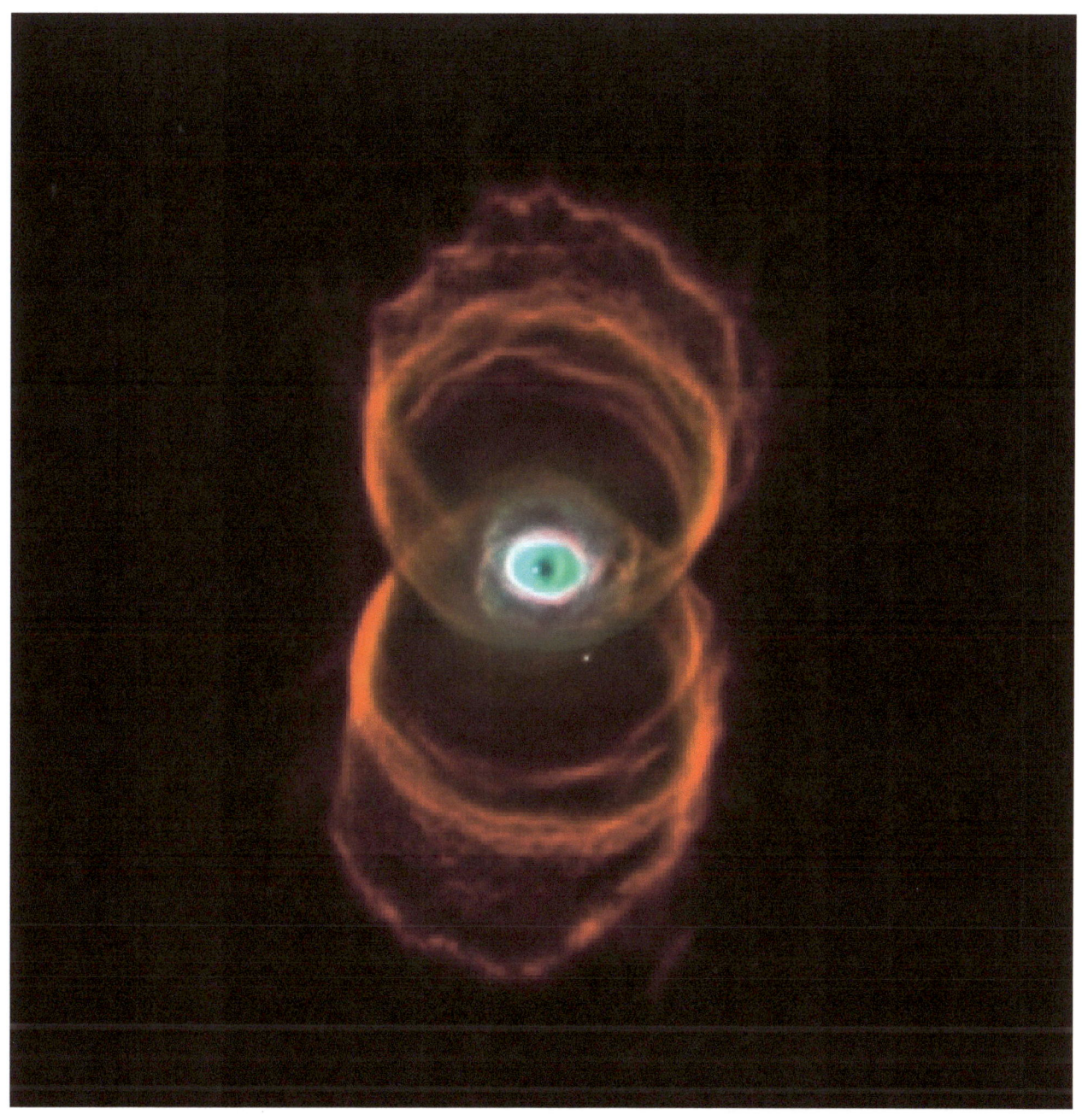

The Hourglass Nebula (MyCn18), a supernova remnant

By NASA, R. Sahai, J. Trauger (JPL), and The WFPC2 Science Team - http://www.spacetelescope.org/images/opo9607a/, Public Domain, https://commons.wikimedia.org/w/index.php?curid=1849193

The pinch takes only a few hours to form. At its centre is a brand new star, quite possibly surrounded by planets with moons.

The Electric Sun

Our Sun is thought to be a big ball of gas with a fusion reactor at its core producing all the heat that radiates from it. However, at closer inspection, the Sun does not look anything like what this theory suggests.

The surface of the Sun, the so called photosphere, looks suspiciously like a liquid. It can produce giant fountains and arches that drip back onto its surface.

When there is a hole in the photosphere that we can peek through, there is nothing to suggest that there is anything going on underneath. Sunspots are black and cool compared to the photosphere.

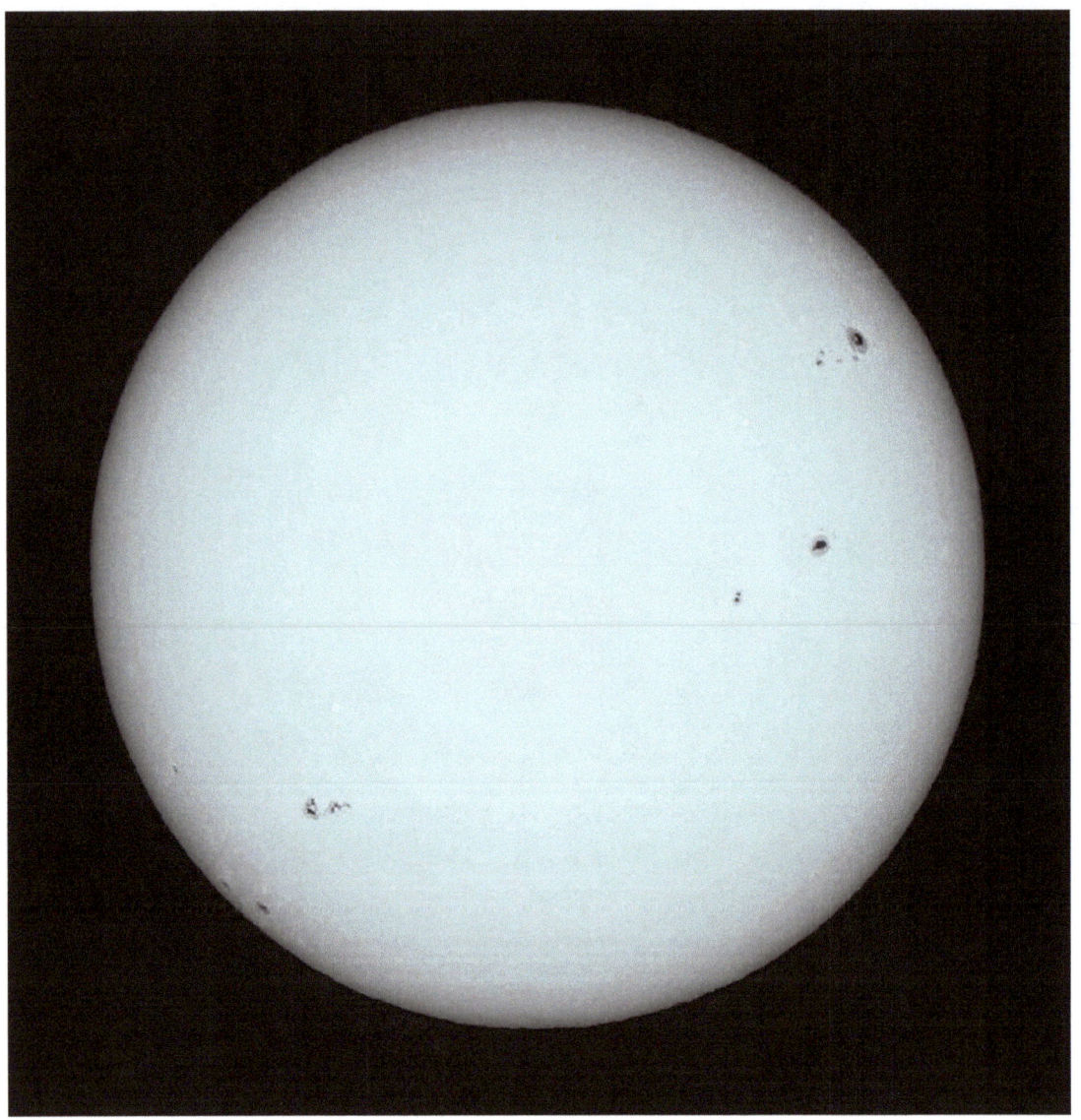

Sun with sunspots

By Geoff Elston - Society for Popular Astronomy, Solar section,
http://www.popastro.com/solar/solarobserving/chapter.php?id_pag=30, CC BY 4.0,
https://commons.wikimedia.org/w/index.php?curid=35976640

The hottest part of the Sun is not close to its core, but in its corona, thousands of miles above its surface.

This is all indicative of electricity. A big difference in voltage potential can accelerate charged particles to enormous speeds, making them increasingly hot as they accelerate towards space. A surface bombarded by charged particles can get so hot that it melts.

It appears then that the current that created the solar system in the first place continues to flow, and that it powers our Sun.

The photosphere is not a gas but liquid rock.

Under the photosphere, where it is relatively cool, stars are solid.

Stars are not made of material significantly different from planets, comets and meteorites. There's no real difference between a star and a planet except for size.

Stars are hotter than planets, simply because they are bigger and therefore the focal point of interstellar currents.

The mistake that has been made regarding the chemical composition of stars is the same that was made for comets. The abundance of water in the tails of comets is due to nuclear fission. Comets are rocky bodies, not dirty snowballs.

The abundance of hydrogen and helium seen in the light spectra of stars is also due to fission, and not due to an abundance of these elements in the star itself.

All the large bodies in our solar system are predominantly made of rock of various kinds. Large planets like Jupiter and Saturn are able to hold onto thick atmospheres, but it is a mistake to think that they have no solid surface. They too have rocky surfaces, just like our Sun.

Stars as Electrical Accelerators

A thing to note about the above hypothesis is that nuclear fission is assumed to take place in the corona of the Sun, and most likely in the chromosphere and photosphere as well.

Sun's corona and chromosphere, visible to the naked eye during a total eclipse

By I, Luc Viatour, CC BY-SA 3.0,
https://commons.wikimedia.org/w/index.php?curid=1107408

Nuclear fission is an exothermic reaction for all materials heavier than iron.

This means that energy is added to the external environment through fission of heavy elements. The Sun is an electrical accelerator. It has a greater output of electric energy than its input.

This in turn goes a long way in explaining where the cosmic currents originate in the first place. The source of the currents that power our Sun is other stars.

Supernovas as Endothermic Heat Sinks

Although extremely hot and bright, supernovas are most likely endothermic. The creation of a solar system pulls energy out of the environment. This energy is used to synthesize an abundance of heavy elements in the objects formed.

Supernova SN 1994D, lower left, outshines its home galaxy

By NASA/ESA, CC BY 3.0, https://commons.wikimedia.org/w/index.php?curid=407520

Life-Cycle of Matter

The overall progression of matter goes from light to heavy through mass condensation.

This makes matter more radioactive, and therefore more readily fissionable. This in turn extends the life of stars. The gradual increase in radioactivity allows stars to function as electrical accelerators for longer.

However, the process of mass condensation cannot go on for ever. At some point, all elements will become radioactive. There will only be hydrogen left, all with enormous protons.

It seems unlikely that this is a sustainable state of matter. Such a situation begs for some correcting mechanism. My guess is that protons have an upper limit to their size, and that the proton itself becomes radioactive once that limit is breached.

Once protons start to collapse on themselves, a chain reaction ensues, and we get a massive burst of gamma-rays.

Bundled up with these gamma-rays, we find electrons and positrons that soon start the process of mass condensation all over again.

This is how quasars are formed. Over time, they become galaxies. The galaxies develop regions of degenerate matter. The matter collapses into radiation, and the cycle repeats.

Life-cycle of a galaxy, from quasar to maturity

The Eternal Universe

Once we accept the idea that electricity, rather than gravity, is the main force in the universe, many things become easier to explain.

We no longer need a super-dense crystal at the core of our planet.

There is no need for dark energy, dark matter, black holes or a big bang.

There is no need for a beginning or an end to the universe.

Instead, we have an eternal universe with no start and no end. Some areas are young. Others are old. Creation and destruction happen continuously and everywhere.

Galaxy cluster IDCS J1426

By ESA/Hubble, CC BY 4.0,
https://commons.wikimedia.org/w/index.php?curid=46299179

Where exactly our region of the universe is in this cycle is hard to say, but I suspect we are somewhere in the middle, perhaps a little closer to the end than the start.

Distance

When it comes to the four physical quantities of distance, time, energy and inertia, it is tempting to treat them as if they are somehow outside of physics. We can imagine a god in the heavens holding a ruler and a clock, distributing energy and bestowing inertia onto matter.

However, that would not be physics, and this is a book about physics, so we must find another way to define these quantities if we are to include them in our model.

The way we can do this is as follows:

To measure distance, we need a ruler of some kind. In our daily lives, the ruler we use is ourselves. We measure everything relative to our own size.

However, when we want to be precise about our measurements, we use a carefully crafted ruler.

Such a ruler is something we can carry around with us. Its length does not change and it does not fly about on its own.

The smallest possible ruler we can make is therefore the electron. Things smaller than an electron moves about at the speed of light and can therefore not be used.

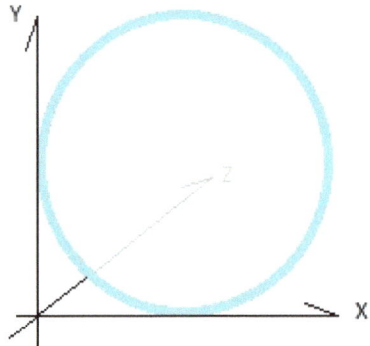

The electron as a three dimensional ruler

Distance is in other word a function of matter in the open state.

Time

To measure time, we need a clock. We all carry our own biological clock inside of us, and that is the clock we normally use. However, when we want to be precise about time, we build ourselves a clock.

The way a clock works is that it takes something that moves at a very predictable speed and make it give off a tick every time it has moved an equally precisely measured distance.

A time unit is always between two ticks. A single tick is not a time unit. That's why we say that clocks go tick-tock or tick-tick.

The smallest possible time unit we can register is in other words a function of the smallest possible ruler and the fastest possible speed.

Since all matter in the closed state moves at the speed of light, we have a ready supply of stuff moving at a very precise speed.

Our smallest possible ruler is the electron.

The smallest possible time unit is therefore the time it takes a photon to cross an electron. If something happens faster than this, the time laps cannot be registered in any way.

An instantaneous event is anything that happens faster than it takes a photon to cross an electron.

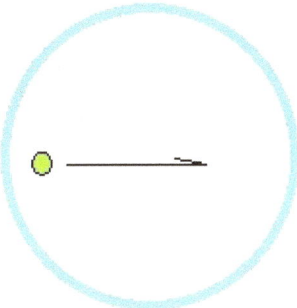

The electron as a clock

Time is in other words a function of matter in the open state combined with matter in the closed state.

We get: Time = Size of matter in the open state / Speed of matter in the closed state

Energy

We have on several occasions in this book noted the relationship between energy and size of particles. Large photons carry more energy than small ones. Electrons carry more energy than photons, etc.

This relationship between size and energy is no coincidence, and I will therefore propose that energy is in fact size of subatomic particles. The energy of matter in the closed state is carried entirely as size.

This is also true for matter in the open state.

What is strange about matter in the open state is that it can move at a variable speed. However, it is not the speed of particles that carry energy. It is their size. When we accelerate a particle in the open state, it grows in size by a tiny bit.

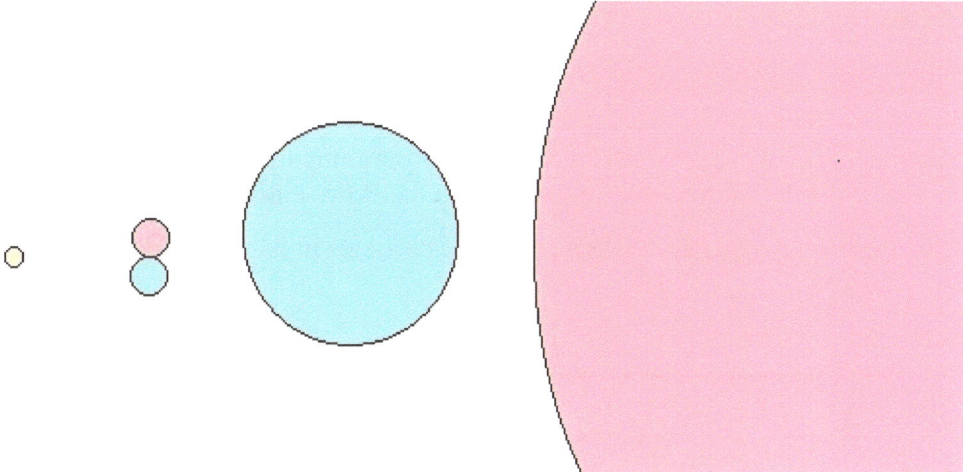

The bigger the particle, the more energy it carries

Energy is in other words entirely a function of state and size of particles. The state is either closed or open, while the size is variable.

Inertia

Inertia is a resistance to change in energy. The more massive something is, the harder we have to push to change its speed or direction.

There is a time delay in the energy transfer.

The fact that inertia only exists in matter in the open state is a huge hint as to its nature.

Energy transfers to and from matter in the closed state happens instantaneously because matter in the closed state is very small. It happens faster than can be registered.

However, the time it takes to transfer energy to and from matter in the open state can be registered. It takes more time to transfer energy onto an electron than it takes for a photon to cross it.

Energy transfers are about readjusting sizes of particles. For matter in the closed state, this happens faster than it takes a photon to cross an electron. For matter in the open state, it takes more time.

For an object to change its speed or direction, all its particles have to change size. For large objects, this requires a lot of energy and time.

Big ship, big inertia

By Wmeinhart - Foto wurde mit einem Panoramaprogramm aus drei Fotos zusammengesetzt, CC BY-SA 3.0, https://commons.wikimedia.org/w/index.php?curid=124261

Inertia is the time delay registered when energy of particles in the open state are changed.

Time and Energy

If energy is stored as size in subatomic particles, and the rate of time is related to the size of these same particles, then time can be expected to slow down for any particle that experiences an increase in energy.

This is because the "electron-clock" is bigger, and therefore slower.

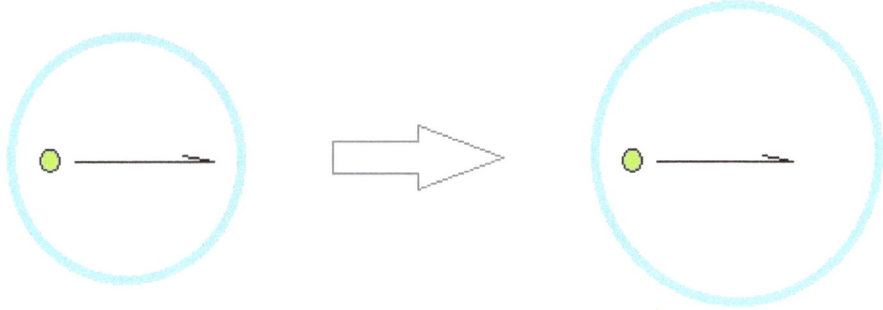

Bigger electron = longer distance for photon to travel = slower time

Observed from outside, the natural processes associated with a speeding atom will appear to slow down.

This is exactly what we find.

Radioactive particles travelling close to the speed of light decay less quickly than the same particles do when stationary.

The Mercury Anomaly

Mercury makes its rounds around the Sun a little faster than predicted by Newton.

This is currently explained using a formula in which time and space is bent. However, this can also be explained using the physics laid out in this book.

The only additional requirement to those already presented is for matter in the open state to be hollow and allowing for the aether of zero-point particles to flow freely into and out of their inside.

With this addition, we get electrical pressure on the inside of electrons and protons. Zero-point neutrinos will bounce about on the inside of open state particles.

The precise size of a particle in the open state is no longer only dependent on the energy it is carrying, but also dependent on the availability of neutrinos.

In regions of space where there is an abundance of neutrinos, particles in the open state will be larger than in regions with relatively fewer neutrinos.

When we combine this with the fact that zero point photons are dielectric, and therefore more abundant close to massive objects, we get that particles in the open state are smaller in such regions.

Zero-point photons close to massive bodies supplant neutrinos, making neutrinos relatively more scarce.

With particles in the open state being smaller closer to massive bodies, we get that our "electron clock" goes faster.

The "electron clock" on Earth is bigger than the one on Mercury

Time on Earth goes slower than time on Mercury, not because time-space is curved, but because particles in the open state are smaller on Mercury than on Earth.

Using a clock on Earth to measure Mercury's orbit, we find that Mercury takes the rounds a little faster than predicted by Newton's formula. However, if we use a clock on Mercury, things will be exactly as predicted. Relative to a clock on Mercury, it is all the other planets that are too slow.

Conclusion

Starting off with three particle quanta and the idea that nothing happens without direct physical interaction, we have successfully laid out a physics in which everything from forces and optics to distances, time and energy are explained.

In this physics, there is no need for any mysterious action at a distance or undetermined state of things. Uncertainty is purely a function of complexity.

Nor is there any need for a curved space-time. Instead, we have particles interacting with other particles.

This is not to say that the physics laid out in this book is how things necessarily must be. Things may be very different. All we have demonstrated is that there are more than one way to explain things.

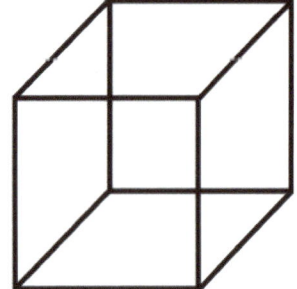 

The Necker cube and Rubin vase can be perceived in more than one way

By Alan De Smet at English Wikipedia - Transferred from en.wikipedia to Commons., Public Domain, https://commons.wikimedia.org/w/index.php?curid=2428267